# Holt Mathematics

## Chapter 14 Resource Book

**HOLT, RINEHART AND WINSTON**
A Harcourt Education Company
Orlando • **Austin** • New York • San Diego • London

Printed in the United States of America



ISBN 0-03-078406-9

6 7 8 170 09 08

# CONTENTS

## Blackline Masters

Date ___________

Dear Family,

In this chapter your child will learn about polynomials. Your child will learn how to identify and simplify polynomials and how to distinguish among monomials, binomials, and trinomials. Your child will also learn how to add and subtract polynomials, multiply a polynomial by a monomial, and multiply two binomials.

Here are some examples of monomials and products that are not monomials.

| | |
|---|---|
| Monomials | $6x$, $n^3$, $2a^3b^2$, 18 |
| Not monomials | $y^{2.5}$, $5^x$, $\frac{6}{n^2}$ |

A monomial has one term, a binomial has two terms, and a trinomial has three terms.

Polynomials can be simplified by adding and subtracting like terms, which have the same variables raised to the same powers.

Like terms

$$2a^2b^3 + 3a^2b^2 - 5a^2b^3$$

Simplify this polynomial by combining like terms.

$3xy + 5xy^2 - 5 + 2xy^2$

$5xy^2 + 2xy^2 + 3xy - 5$ Arrange in descending order.

$\boxed{5xy^2} + \boxed{2xy^2} + 3xy - 5$ Identify like terms.

$7xy^2 + 3xy - 5$ Combine coefficients.

To add or subtract polynomials, your child will combine like terms.

$(3a^2 + 2a + 5) + (6a^2 + 3a + 2)$

$$\begin{array}{r} 3a^2 + 2a + 5 \\ +\ 6a^2 + 3a + 2 \\ \hline 9a^2 + 5a + 7 \end{array}$$

Place like terms in columns.

Combine like terms.

To multiply a polynomial by a monomial, your child will use the Distributive Property.

$3x^2(2x^3 - 3x - 5)$

$3x^2 \cdot 2x^3 - 3x^2 \cdot 3x - 3x^2 \cdot 5$ Use the Distributive Property.

$6x^5 - 9x^3 - 15x^2$

To multiply two binomials, your child will multiply each term of one binomial by each term in the other binomial.

The product can be remembered as FOIL: First terms, Outer terms, Inner terms, and Last terms.

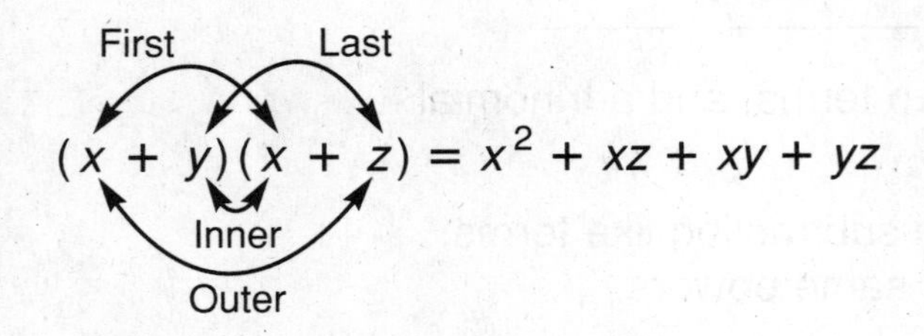


$(a + 2b)(3a + 5b)$

$(a + 2b)(3a + 5b)$

$3a^2 + 5ab + 6ab + 10b^2$

$3a^2 + 11ab + 10b^2$ Combine like terms.

For additional resources, visit go.hrw.com and enter the keyword MT7 Parent.

Name ______________________ Date ____________ Class ____________

LESSON 14-1

# Practice A

## *Polynomials*

**Determine whether each expression is a monomial.**

**1.** $-x^3$ ____________

**2.** $4xy^9$ ____________

**3.** $\frac{8}{q^3}$ ____________

**4.** $3.5r^{\frac{1}{2}}$ ____________

**5.** $a^{2.2}$ ____________

**6.** $\frac{7}{9}xyz$ ____________

**Classify each expression as a monomial, a binomial, a trinomial, or not a polynomial.**

**7.** $-8.9x + 6x^5$ ____________

**8.** $\frac{1}{18}a^8a^2$ ____________

**9.** $x^8 + x + \frac{1}{x}$ ____________

**10.** $-7r^4 + s^3$ ____________

**11.** $m^{15} - m + \frac{1}{9}$ ____________

**12.** $-7.55r^{75}$ ____________

**Find the degree of each polynomial.**

**13.** $\frac{7}{2} - x$ ____________

**14.** $a + a^2 + a^3$ ____________

**15.** $7w - 6w + 3w$ ____________

**16.** $-9p - 9p - 9p^3$ ____________

**17.** $y^9 + 10y^8 - 7y$ ____________

**18.** $1{,}055 + \frac{4}{5}k - k^7$ ____________

**19.** The volume of a box with width $x$, length $2x + 1$, and height $2x - 2$ is given by the trinomial $4x^3 - 2x^2 - 2x$. What is the volume of the box if its width is 5 meters?

________________________________________

Name ______________________ Date __________ Class __________

LESSON **14-1**

# Practice B
## *Polynomials*

**Determine whether each expression is a monomial.**

**1.** $-135x^5$

**2.** $2.4x^3y^{19}$

**3.** $\frac{2p^2}{q^3}$

**4.** $3r^{\frac{1}{2}}$

**5.** $43a^2b^{6.1}$

**6.** $\frac{7}{9}x^2yz^5$

**Classify each expression as a monomial, a binomial, a trinomial, or not a polynomial.**

**7.** $-8.9xy + \frac{6}{y^5}$

**8.** $\frac{9}{8}ab^8c^2d$

**9.** $x^8 + x + 1$

**10.** $-7pq^{-2}r^4$

**11.** $5n^{15} - 9n + \frac{1}{3}$

**12.** $r^8 - 5.5r^{75}$

**Find the degree of each polynomial.**

**13.** $7 - 14x$

**14.** $5a + a^2 + \frac{6}{7}a^3$

**15.** $7w - 16u + 3v$

**16.** $9p - 9q - 9p^3 - 9q^2$

**17.** $z^9 + 10y^8 - x$

**18.** $100{,}050 + \frac{4}{5}k - k^4$

**19.** The volume of a box with height $x$, length $x - 1$, and width $2x + 2$ is given by the binomial $2x^3 - 2x$. What is the volume of the box if its height is 4 feet?

______________________________

**20.** The trinomial $-16t^2 + 32t + 32$ describes the height in feet of a ball thrown upward after $t$ seconds. What is the height of the ball $\frac{5}{8}$ seconds after it was thrown?

______________________________

Name ______________________ Date __________ Class __________

LESSON 14-1

# Practice C

## Polynomials

**Determine whether each expression is a monomial.**

**1.** $-1.35a^{135}b^{12}c^{3}$

**2.** $2.4x^{\frac{3}{2}}y^{19}$

**3.** $\frac{2p^2q^5}{pq^3}$

______________ ______________ ______________

**Classify each expression as a monomial, a binomial, a trinomial, or not a polynomial.**

**4.** $-0.9x + \frac{6}{y^5} + 1$

**5.** $\frac{8ab^8c^2d}{ad}$

**6.** $-8x^8 + x + 15x^6$

______________ ______________ ______________

**7.** $-7h^{-2}k$

**8.** $5m^{15} - 9n^9 - 0.03$

**9.** $0.6t^6 - 75g^0t^{75}$

______________ ______________ ______________

**Find the degree of each polynomial.**

**10.** $50 - 1.024s^6$

**11.** $5u^2 + u + \frac{6}{5}u^3$

**12.** $\frac{7}{1}m - 2.05n + 3p$

______________ ______________ ______________

**13.** $9(p - \frac{2}{3}q - 9p^3)$

**14.** $(z^9)^0 + 10y^{89} - x$

**15.** $\frac{4}{11}h^4k - k^4$

______________ ______________ ______________

**16.** The area of a rectangle with length $x$ and width $2x - 3$ is given by the binomial $2x^2 - 3x$. What is the area of the rectangle if its length is 5 yards?

________________________________________

**17.** The height in feet of a ball thrown straight up into the air from $s$ feet off the ground at velocity $v$ after $t$ seconds is given by the polynomial $-16t^2 + vt + s$. Find the height of a ball thrown from a 20 ft platform at 150 ft/s after 6 seconds.

________________________________________

Name ______________________ Date ___________ Class ___________

LESSON **14-1**

# Reteach

## *Polynomials*

Expressions such as $2x$ and $4y^2$ are called **monomials.** A monomial has only one term. Monomials do <u>not</u> have fractional exponents, negative exponents, variable exponents, roots of variables, or variables in a denominator.

**Determine whether each expression is a monomial.**

**1.** $3x - 5$ **2.** $-9a^4$ **3.** $21m^{0.5}$ **4.** $7m^3n^2$

__________ __________ __________ __________

A monomial or a sum or difference of monomials is called a **polynomial.** Polynomials can be classified by the number of terms. A monomial has 1 term, a **binomial** has 2 terms, and a **trinomial** has 3 terms.

**Classify each expression as a monomial, a binomial, a trinomial, or not a polynomial.**

**5.** $7y + 3x^2 + 5$ **6.** $6y + \sqrt{x}$

__________ __________

**7.** $m^2n$ **8.** $-6a + 2b^4$

__________ __________

The degree of a polynomial is the degree of the term with the greatest degree. The **degree** of a term is the greatest value of a variable's exponent.

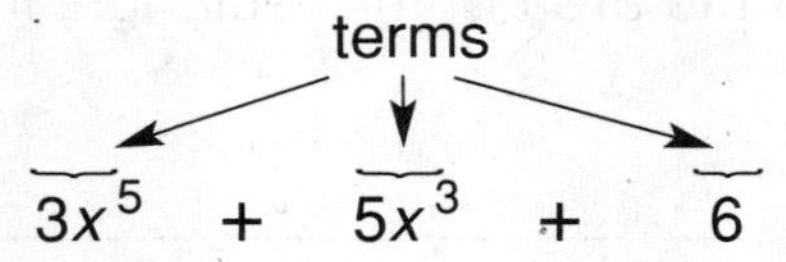


**5th degree** 3rd degree 0 degree

The above polynomial is a 5th degree trinomial.

**Find the degree of each polynomial.**

**9.** $5x + 3x^3 + 2x^2$ **10.** $-3m^4 + m^2 + 2$ **11.** $4y + 2y^3 + y^5$ **12.** $7a^2 + 8a$

__________ __________ __________ __________

Name ______________________________ Date ____________ Class ____________

LESSON 14-1

# Challenge

## Polynomial Means "Many Names"

How many different ways can you name the expression $3x^3 + 4x + 7$? You can call it a polynomial because it is a sum of monomials, or a trinomial because it is a polynomial with 3 terms. You can also call it a polynomial of third degree because the term with the greatest degree has degree 3. But did you know that some polynomials have even more names?

**For each polynomial described in the chart below, write an example of a monomial, a binomial, and a trinomial to match its special degree name.**

| Degree | Names | Monomial | Binomial | Trinomial |
|---|---|---|---|---|
| 1 | Linear or Monic | | | |
| 2 | Quadratic | | | |
| 3 | Cubic | | | |
| 4 | Quartic | | | |
| 5 | Quintic | | | |
| 6 | Sextic or Hexic | | | |
| 7 | Septic or Heptic | | | |
| 8 | Octic | | | |
| 9 | Nonic | | | |
| 10 | Decic | | | |

**Extra:**

What degree of polynomial is named "Hectic"? ____________

Evaluate the polynomial $9x^2 + 15x + 4x^3 + x$ for $x = 2$. ____________

Name ______________________ Date ____________ Class ____________

LESSON 14-1

# Problem Solving

## *Polynomials*

**The table below shows expressions used to calculate the surface area and volume of various solid figures where *s* is side length, *l* is length, *w* is width, *h* is height, and *r* is radius.**

**Solid Figure Polynomials**

| Solid Figure | Surface Area | Volume |
|---|---|---|
| Cube | $6s^2$ | $s^3$ |
| Rectangular Prism | $2lw + 2lh + 2wh$ | $lwh$ |
| Right Cone | $\pi rl + \pi r^2$ | $\pi r^2 h$ |
| Sphere | $4\pi r^2$ | $\frac{4}{3}\pi r^3$ |

1. List the expressions that are trinomials.

______________________

2. What is the degree of the expression for the surface area of a sphere?

______________________

3. A cube has side length of 5 inches. What is its surface area?

______________________

4. If you know the radius and height of a cone, you can use the expression $(r^2 + h^2)^{0.5}$ to find its slant height. Is this expression a polynomial? Why or why not?

______________________

______________________

5. If a sphere has a radius of 4 feet, what is its surface area and volume? Use $\frac{22}{7}$ for pi.

______________________

**Circle the letter of the correct answer.**

6. Which statement is true of all the polynomials in the volume column of the table?

**A** They are trinomials

**B** They are binomials.

**C** They are monomials.

**D** None of them are polynomials.

7. The height, in feet, of a baseball thrown straight up into the air from 6 feet above the ground at 100 feet per second after *t* seconds is given by the polynomial $-16t^2 + 100t + 6$. What is the height of the baseball 4 seconds after it was thrown?

**F** 150 feet

**G** 278 feet

**H** 342 feet

**J** 662 feet

Name ______________________ Date __________ Class __________

CHAPTER **14-1**

# Reading Strategies

## *Focus on Vocabulary*

A **monomial** is a number or a product of numbers and variables with exponents that are whole numbers. A **polynomial** is a monomial or the sum or difference of monomials. This chart can help you understand polynomials.

| **Examples** | **Classifying Polynomials by the Number of Terms** | **Classifying Polynomials by their Degree** |
|---|---|---|
| $3a^2b \rightarrow$ monomial | monomial = 1 term | $15x + 9x^4 - 4x^2 + 20$ (with $9x^4$ circled) |
| $\frac{4x^2}{y^3} \rightarrow$ not a monomial | binomial = 2 terms | highest power = degree = 4 |
| $6x2y - y3 \rightarrow$ polynomial | trinomial = 3 terms | $-13x^4 + x - 2x^6$ (with $2x^6$ circled) |
| $x^3 + 2y^{\frac{1}{2}}$ | | highest power = degree = 6 |

**Use the information in the chart to answer the following questions.**

1. What is a monomial?

______________________________________________

______________________________________________

2. What is a polynomial?

______________________________________________

3. Explain why $x^3 + 2y^{\frac{1}{2}}$ is not a polynomial.

______________________________________________

4. Rewrite $\frac{4x^2}{y^3}$ so that it does not have a variable in the denominator.
   Use your answer to explain why $\frac{4x^2}{y^3}$ is not a polynomial.

______________________________________________

______________________________________________

5. Classify the polynomial $-2x^2 + 5x + 7$ by the number of terms.

______________________________________________

6. Classify the polynomial in Exercise 5 by its degree.

______________________________________________

Name ______________________ Date ____________ Class ____________

LESSON 14-1

# Puzzles, Twisters & Teasers

## *Puzzle It Out!*

**Complete the crossword puzzle using terms you learned in this lesson.**

**Across**

1. A polynomial with three terms is a ____________.
5. An expression is not a polynomial if it has a variable in the ____________.
7. A letter that represents an unknown quantity is a ____________.
8. The exponents in the terms of a polynomial must be ____________ numbers.
9. A polynomial with two terms is a ____________.

**Down**

2. The simplest type of polynomial is a ____________.
3. A number that precedes a variable is called the ____________ of the variable.
4. One monomial or the sum or difference of monomials is a ____________.
5. You look at the exponents in a polynomial to determine its ____________.
6. A trinomial is a polynomial with three ____________.

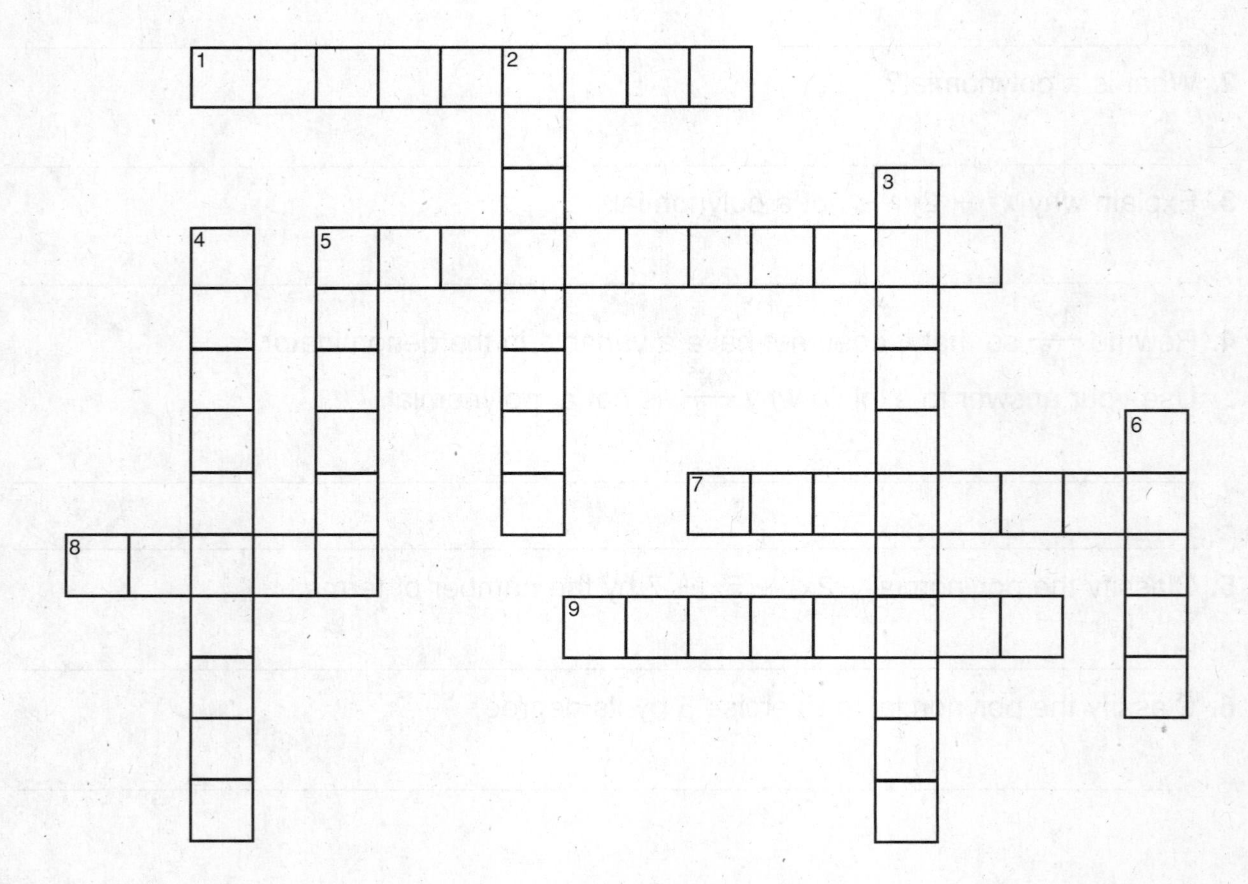

Name ______________________ Date __________ Class __________

LESSON **14-2**

# Practice A

## *Simplifying Polynomials*

**Identify the like terms in each polynomial.**

1. $x^2 - x + 3x^2 + 6x - 1$

   ______________________

2. $2t^2 + 7 - s^3 - t^2$

   ______________________

3. $2e^2 - 2 - 8e^2 + 3e - 2e^2$

   ______________________

4. $2k - 3k^2 + 3k^2 - k$

   ______________________

5. $4ab^2 - 2b + 10 - ab^2$

   ______________________

6. $6z^2 + yz^2 + z^2 - 3yz^2$

   ______________________

7. $3g^2h^2 + 2g^2h - 7 + g^2h - 5g^2h^2$

   ______________________

8. $m^3n^2 + 6 - m^2n^2 - 2m^3n^2 + 4m^2n^2$

   ______________________

**Simplify.**

9. $k^2 - k + 2k^2 - 2k$

   ______________________

10. $3x - 2 + 4x^2 + 6 - x^2$

    ______________________

11. $7r^2 - 7 - 23 - 3r^2$

    ______________________

12. $v^4 + v^2 + 2v^2 - 1$

    ______________________

13. $2(x + 2y)$

    ______________________

14. $3y + 2w - 7y + 5w$

    ______________________

15. $4(2r - 2) - 10$

    ______________________

16. $7 - w - 2 + 3w + 7$

    ______________________

17. The height of a ball dropped off of a roof after $t$ seconds is given by the polynomial $-4(4t^2 - 10)$. Use the Distributive Property to write an equivalent expression.

    ______________________

Name ______________________ Date ____________ Class ____________

LESSON 14-2

# Practice B
## *Simplifying Polynomials*

**Identify the like terms in each polynomial.**

**1.** $x^2 - 8x + 3x^2 + 6x - 1$

______________________

**2.** $2c^2 + d^3 + 3d^3 - 2c^2 + 6$

______________________

**3.** $2x^2 - 2xy - 2y^2 + 3xy + 3x^2$

______________________

**4.** $2 - 9x + x^2 - 3 + x$

______________________

**5.** $xy - 5x + y - x + 10y - 3y^2$

______________________

**6.** $6p + 2p^2 + pq + 2q^3 - 2p$

______________________

**7.** $3a + 2b + a^2 - 5b + 7a$

______________________

______________________

**8.** $10m - 3m^2 + 9m^2 - 3m - m^3$

______________________

______________________

**Simplify.**

**9.** $2h - 9hk + 6h - 6k$

______________________

**10.** $9(x^2 + 2xy - y^2) - 2(x^2 + xy)$

______________________

**11.** $7qr - q^2r^3 + 2q^2r^3 - 6qr$

______________________

**12.** $8v^4 + 3v^2 + 2v^2 - 16$

______________________

**13.** $3(x + 2y) + 2(2x - 3y)$

______________________

**14.** $7(1 - x) + 3x^2y + 7x - 7$

______________________

**15.** $6(9y + 1) + 8(2 - 3y)$

______________________

**16.** $a^2b - a^2 + ab^2 - 3a^2b + ab$

______________________

**17.** A student in Tracey's class created the following expression: $y^3 - 3y + 4(y^2 - y^3)$. Use the Distributive Property to write an equivalent expression.

______________________________________________

Name ______________________ Date ____________ Class ____________

LESSON 14-2

# Practice C

## Simplifying Polynomials

**Identify the like terms in each polynomial.**

**1.** $x^2 - 18x + 3.5x^2 + 16x - 0.3$

**2.** $s^3t + s^3 + 3s^3t - t^2 + 4st + 2s^3$

**3.** $\frac{1}{3}g + \frac{1}{2}h + g^2 - \frac{5}{2}h$

**4.** $\frac{1}{2}p^2 - 4p + 1 - p + \frac{9}{2}$

**5.** $12e^2 - 21ef - 8f^2 + 3ef - 22e^2$

**6.** $2cd - c^2d^2 + 9c - 3c^2d^2 + c$

**7.** $\frac{1}{5}mn - 3m^2n + \frac{3}{5}mn - mn^2$

**8.** $6m - m^5 + 2m^3 - 2m$

**Simplify.**

**9.** $2k - 9kn - 2k - 2k + 5(n - k)$

**10.** $9(x^2 + xy - y^2) + \frac{1}{9}(x^2 - 9xy)$

**11.** $\frac{7}{2}qr - q^2r^3 + \frac{1}{2}(q^2r^3 - 3qr)$

**12.** $\frac{8}{3}v^4 + v^2 + \frac{1}{2}(v^2 - 4)$

**13.** $4(x + 2y) - 2(2x - 3y) - 2$

**14.** $7(1 - w) + 7(-2w + 3w^2 + 1) + 7$

**15.** $2(6 + 9q) - (6q - 9q^2 + 6q)$

**16.** $abc + a^2 + a^3b^2c - a^2 + abc$

**17.** Mark's math teacher wrote the following expression on the board: $\frac{1}{2}\left(8 - 4y - \frac{1}{2}y^2\right) + \frac{1}{8}$. Use the Distributive Property to write an equivalent expression. ____________

Name ______________________ Date __________ Class __________

LESSON 14-2

# Reteach

## *Simplifying Polynomials*

You can simplify a polynomial by combining like terms. Like terms have the same variables raised to the same powers. All constants are like terms.

$$9 + 6y^3 - 8 + 7x^2y^3 + 3x^2y^3$$

like terms ($9$ and $-8$)   like terms ($7x^2y^3$ and $3x^2y^3$)

$7x^2y^3$ and $3x^2y^3$ both have the variable $x$ raised to the 2nd power and the variable $y$ raised to the 3rd power. Therefore, they are like terms.

**Identify the like terms in each polynomial**

**1.** $m + 3m^2 - 2m + 6 + 2m^2$

______________________

**2.** $b - a^2b^2 - 2 + a^2 + 2a^2b^2$

______________________

**3.** $x^3 + 2 + 4x^3 - 9 + x$

______________________

**4.** $9 + 4dg^2 + 4 + 6dg^2 + d^2$

______________________

To simplify a polynomial, combine like terms. To combine like terms, add or subtract the coefficients. The variables and the exponents
do not change.

$7x^2y^3 - 6y^3 + 3x^2y^3$ (like terms $7x^2y^3$ and $3x^2y^3$ circled)

$\mathbf{7}x^2y^3 + 6y^3 + \mathbf{3}x^2y^3$   Identify like terms.

$\mathbf{10}x^2y^3 + 6y^3$   Combine coefficients of like terms.

$7 + 3 = 10$

**Simplify.**

**5.** $8a + 3ab^2 + 3a + 2ab^2$

______________________

**6.** $x^3 + 1 + 2x^3 + 3xy^2 - 3$

______________________

**7.** $y^4 + 2x^2y^3 - 3x^2 + 2y^4$

______________________

Name ______________________ Date __________ Class __________

# Reteach

## 14-2 Simplifying Polynomials (continued)

To simplify some polynomials, you first need to use the Distributive Property.

Distributive Property: $a(b + c) = a \cdot b + a \cdot c$

Simplify the polynomial.

$3(4y^2 + x^2) + 5x^2$

$\mathbf{3}(4y^2 + x^2) + 5x^2 = (\mathbf{3} \cdot 4y^2) + (\mathbf{3} \cdot x^2) + 5x^2$ Apply the Distributive Property.

$= 12y^2 + 3x^2 + 5x^2$

$= 12y^2 + 3x^2 + 5x^2$ Combine like terms.

like terms

$= 12y^2 + 8x^2$

So, $3(4y^2 + x^2) + 5x^2$ simplified is $12y^2 + 8x^2$.

**Simplify.**

**8.** $2(2a^2 - b^2) + 3a^2$

______________________

**9.** $4(3x^2y + 2x) + 3x^2 + 5x$

______________________

**10.** $-6mn + 3(m^4 + 3mn) - 2n^2$

______________________

**11.** $8(y - 5) + 3y + 6y^2$

______________________

**12.** $2(x^2 + 3x - 6) + 5(y + 2)$

______________________

**13.** $4(xy^2 + 2y + 3) + 2(x^2 - 2xy^2)$

______________________

Name ______________________ Date __________ Class __________

LESSON 14-2

# Challenge

## *Coming To Terms*

**For each polynomial, use the simplified polynomial to find the missing term.**

| | Polynomial | Simplified Polynomial | Missing Term |
|---|---|---|---|
| **1.** | $7x^2 - 4x + \underline{?} + 3x - 5$ | $5x^2 - x - 5$ | |
| **2.** | $6(9x + \underline{?})$ | $54x + 18$ | |
| **3.** | $12 + 2m + 3m^4 - 8m^2 + 5 + \underline{?} - 7m^2$ | $3m^4 - 15m^2 + 6m + 17$ | |
| **4.** | $\underline{?}\,(2b^2 - 9b) + 4b^2 + 11b$ | $14b^2 - 34b$ | |
| **5.** | $7(x^3 + 3x) - 5x^3 + \underline{?}$ | $2x^3 + 11x$ | |
| **6.** | $4ab + \underline{?} + 3a^2b^2 + 2ab - 8$ | $2a^2b^2 + 6ab - 8$ | |
| **7.** | $\underline{?} + 7 + 10w^2 - 4w^2 + 5w - 2$ | $6w^2 + 4w + 5$ | |
| **8.** | $2(t - 7) + \underline{?} - 2t^2$ | $-2t^2 + 14t - 14$ | |
| **9.** | $3hk - h^2k + hk^2 + 4hk + \underline{?} + 3hk^2$ | $4hk^2 - 3h^2k + 7hk$ | |
| **10.** | $4y^3 + 6y - 7y^2 + 2y^3 + \underline{?}$ | $6y^3 - 6y^2 + 6y$ | |
| **11.** | $\underline{?} - n^3 - \frac{1}{4}n^4 + \frac{1}{2}n^3 - \frac{1}{3}n^3$ | $\frac{1}{4}n^4 - \frac{5}{6}n^3$ | |
| **12.** | $3(12v^3 + \underline{?} + 2v^3) + v^3$ | $46v^3$ | |
| **13.** | $1.5pq^3 + 0.7p^2q + \underline{?} + 2.4p^2q$ | $1.5pq^3 + 3.1p^2q - 3.8pq$ | |
| **14.** | $-9(\underline{?} + 8w) + 4(-2w^2 + 18w)$ | $w^2$ | |

Name ______________________ Date __________ Class __________

LESSON **14-2**

# Problem Solving

## *Simplifying Polynomials*

**Write the correct answer.**

**1.** The area of a trapezoid can be found using the expression $\frac{h}{2}(b_1 + b_2)$ where $h$ is height, $b$ is the length of base$_1$, and $b_2$ is the length of base$_2$. Use the Distributive Property to write an equivalent expression.

______________________

**2.** The sum of the measures of the interior angles of a polygon with $n$ sides is $180(n - 2)$ degrees. Use the Distributive Property to write an equivalent expression, and use the expression to find the sum of the measures of the interior angles of an octagon.

______________________

**3.** The volume of a box of height $h$ is $2h^4 + h^3 + h^2 + h^2 + h$ cubic inches. Simplify the polynomial and then find the volume if the height of the box is 3 inches.

______________________

______________________

**4.** The height, in feet, of a rocket launched upward from the ground with an initial velocity of 64 feet per second after $t$ seconds is given by $16(4t - t^2)$. Write an equivalent expression for the rocket's height after $t$ seconds. What is the height of the rocket after 4 seconds?

______________________

______________________

**Circle the letter of the correct answer.**

**5.** The surface area of a square pyramid with base $b$ and slant height $l$ is given by the expression $b(b + 2l)$. What is the surface area of a square pyramid with base 3 inches and slant height 5 inches?

**A** 13 square inches

**B** 19 square inches

**C** 39 square inches

**D** 55 square inches

**6.** The volume of a box with a width of $3x$, a height of $4x - 2$, and a length of $3x + 5$ can be found using the expression $3x(12x^2 + 14x - 10)$. Which is this expression, simplified by using the Distributive Property?

**F** $36x^2 + 42x - 30$

**G** $15x^3 + 17x^2 - 7x$

**H** $36x^3 + 14x - 10$

**J** $36x^3 + 42x^2 - 30x$

Name ________________________ Date ____________ Class ____________

LESSON 14-2

# Reading Strategies

## *Understand Vocabulary*

**Like terms** in a polynomial have the same variables raised to the same powers. After identifying like terms, you can combine them to simplify the polynomial.

| **Identifying Like Terms** | **Simplifying Polynomials by Combining Like Terms** |
|---|---|
| $7b + 4b^2 + 5 + 3b - b^2$ | Add or subtract the coefficients of like terms. |
| $7b$ and $3b$ are like terms. | Add the coefficients, 7 and 3, and keep the variable the same.<br>$7b + 3b = 10b$ |
| $4b^2$ and $-b^2$ are like terms. | Subtract the coefficients, 4 and $-1$, and keep the variable the same.<br>$4b^2 - b^2 = 3b^2$<br>$7b + 4b^2 + 5 + 3b - b^2 = 10b + 3b^2 + 5$ |

**Answer each question.**

1. Explain why $4b^2$ and $-b^2$ are like terms.

_______________________________________________

2. Explain why $4x^3y^2$ and $15x^2y^2$ are not like terms.

_______________________________________________

3. Describe a monomial that would be a like term for $10xy$. Then give an example.

_______________________________________________

4. Identify the coefficient of each term in the polynomial $3c - 2c^2 + 10$.

_______________________________________________

5. When you combine like terms, what do you do to the coefficients, and to the variables, and to the exponents of the terms?

_______________________________________________

6. In the polynomial $7x^3 - 4x^2 + x + 6$, what is the coefficient of $x$?

_______________________________________________

Name ________________ Date __________ Class __________

LESSON 14-2

# Puzzles, Twisters & Teasers

## *Fix It!*

**A magic square has the same sum for every column, row and diagonal. The sum is called the magic sum. In the magic square below, one of the polynomials is incorrect. Find the magic sum. Then, find the polynomial that does not fit, circle it, and FIX IT!**

| | | |
|---|---|---|
| $4x^2 - 2$ | $9x^2 - 7$ | $2x^2 - 6$ |
| $3x^2 - 9$ | $5x^2 - 5$ | $7x^2 - 1$ |
| $8x^2 - 4$ | $10x^2 - 2$ | $6x^2 - 8$ |

Magic sum ________________

Incorrect polynomial ________________

Corrected polynomial ________________

Name ______________________ Date __________ Class __________

LESSON **14-3**

# Practice A

## *Adding Polynomials*

**Add.**

**1.** $(3a + 3) + (5a + 2)$

______________________

**2.** $(4x - x^2 - 9) + (2 - 2x + 6x^2)$

______________________

**3.** $(5y + 2y^2 + 3) + (-5y^2 + 2y)$

______________________

**4.** $(3k^3 + 7k^2 + k) + (k^3 - 2k^2 - 10)$

______________________

**5.** $(2wv - 2w^2v) + (7wv^2 + 6w^2v) + (-3wv + wv^2)$

______________________

**6.** $(b^2c^2 - b^2c + 3bc) + (b^2c^2 - bc + 1) + (2b^2c - bc - 1)$

______________________

**7.** $(e^2 + 3e + 2) + (9 - 6e + e^2) + (9e - 2 - e^2) + (4e^2 - 7e - 8)$

______________________

**8.** $(f^4 - 2f + f^3 - 4) + (3f^3 + 3) + (f^4 - f) + (3 - f^3 + f^4 - f) + (4 - f)$

______________________

**9.** Micah is putting a frame of width $w$ around an 18-inch by 24-inch poster. Find an expression for the length of framing material she needs.

______________________

**10.** Each side of an equilateral triangle has length $w + 3$. Each side of a square has length $2w + 2$. Write an expression for the sum of the perimeter of the equilateral triangle and the perimeter of the square.

______________________

Name ______________________ Date __________ Class __________

LESSON 14-3

# Practice B

## Adding Polynomials

**Add.**

**1.** $(a^2 + a + 3) + (15a^2 + 2a + 9)$

______________________

**2.** $(5x + 2x^2) + (3x - 2x^2)$

______________________

**3.** $(mn - 10 + mn^2) + (5 + 3mn - 4mn^2)$

______________________

**4.** $(7y^2z + 9 + yz^2) + (y^2z - 2yz^2)$

______________________

**5.** $(s^3 + 3s - 3) + (2s^3 + 9s - 2) + (s - s^3)$

______________________

**6.** $(6wv - 4w^2v + 7wv^2) + (5w^2v - 7wv^2) + (wv^2 - 5wv + 6w^2v)$

______________________

**7.** $(6b^2c^2 - 4b^2c + 3bc) + (9b^2c^2 - 4bc + 12) + (2b^2c - 3bc - 8)$

______________________

**8.** $(7e^2 + 3e + 2) + (9 - 6e + 4e^2) + (9e + 2 - 6e^2) + (4e^2 - 7e + 8)$

______________________

**9.** $(f^4g - fg^3 + 2fg - 4) + (3fg^3 + 3) + (4f^4g - 5fg) + (3 - 12fg^3 + f^4g)$

______________________

**10.** Six blocks of height $4h + 4$ each and 3 blocks of height $8 - 2h$ each are stacked on top of each other to form one big tower. Find an expression for the overall height of the tower.

______________________

Name ______________________ Date __________ Class __________

LESSON 14-3

# Practice C
## Adding Polynomials

**Add.**

**1.** $(3a^2 + 3a + 3) + (15a^2 + 2a - 9)$

______________________

**2.** $(3g^2 + 2hg) + (5g^2 + hg - 9h^2g)$

______________________

**3.** $(5x - 4y + 6y^2) + (-5y^2 + 2y - 3x)$

______________________

**4.** $(m^2 + 3m - 2) + (4m^2 + m - 9)$

______________________

**5.** $\left(\frac{3}{2}s^3 + \frac{1}{6}s^2 + \frac{1}{3}t\right) + \left(\frac{1}{2}s^3 + \frac{1}{3}t\right)$

______________________

**6.** $(6wv - 4w^2v + 7wv^2) + (5w^2v - 7wv^2) + (wv^2 - 5wv + 6w^2v) - 7w^2v$

______________________

**7.** $(6b^2c^2 - 11b^2c + 3.25bc) + (9b^2c^2 - 45bc + 12) + (2b^2c - 3.25bc - 8)$

______________________

**8.** $\left(\frac{1}{7}e^2 + 3e + 2\right) + \left(9 - 6e + \frac{1}{4}e^2\right) + \left(9e - 2 + \frac{6}{7}e^2\right) + (4e^2 - 7e - 8)$

______________________

**9.** $\left(\frac{1}{2}f^4g - 2fg^3 + \frac{1}{2}fg - 4\right) + (3fg^3 + 3) + \left(\frac{2}{4}f^4g - \frac{5}{2}fg\right) + (3 - 12fg^3 + f^4g)$

______________________

**10.** The area of Kenny's back yard is $2h^2 - 9h + 8$ square yards. The area of Kenny's front yard is $h^2 - 38$ square yards. Write an expression for the total area of Kenny's back and front yards.

______________________

Name ______________________ Date __________ Class __________

LESSON 14-3

# Reteach

## *Adding Polynomials*

Adding polynomials is like simplifying polynomials.
You can regroup the terms and then combine like terms. Or you can place the polynomials in columns and then combine like terms.

Find an expression for the perimeter of the triangle below.

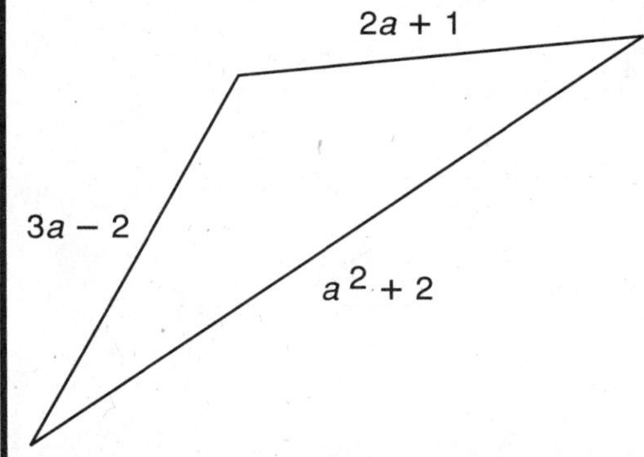


To find an expression for the perimeter, add the polynomials.

$(2a + 1) + (3a - 2) + (a^2 + 2)$

like terms like terms

Place like terms in columns and combine them.

$$\begin{array}{r} 2a + 1 \\ 3a - 2 \\ + a^2 \quad\quad + 2 \\ \hline a^2 + 5a + 1 \end{array}$$

So, an expression for the perimeter of the triangle is $a^2 + 5a + 1$.

**Add.**

**1.** $(3x^2 + 3xy^3 + 5y + 2) + (4xy^3 - 3y)$

$$\begin{array}{r} 3x^2 + 3xy^3 + 5y + 2 \\ + \quad\quad 4xy^3 - 3y \quad\quad \\ \hline \end{array}$$

________________________

**2.** $(4a^2b - 3a^2 + 3b) +$
$(6a^2b + 4ab - 2b)$

$$\begin{array}{r} 4a^2b - 3a^2 \quad\quad\quad + 3b \\ + 6a^2b \quad\quad + 4ab - 2b \\ \hline \end{array}$$

________________________

**3.** $(4mn + 5n^3 + 3n) + (3m^2 + 5n)$

________________________

**4.** $(-5r^3 + 2r + 7) + (2r^3 + 4r^2 - 6r + 1)$

________________________

Name ______________________ Date ____________ Class ____________

LESSON 14-3

## Challenge

### *Perimeters*

**Write a polynomial for the perimeter of each figure. Simplify each polynomial.**

**1.**

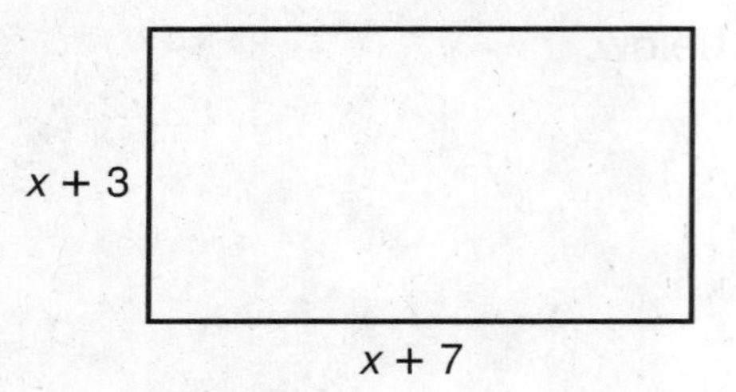


Perimeter: ______________

**2.**

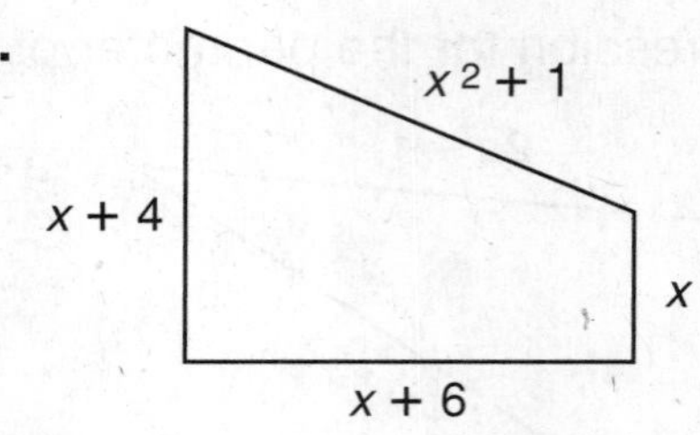


Perimeter: ______________

**3.**

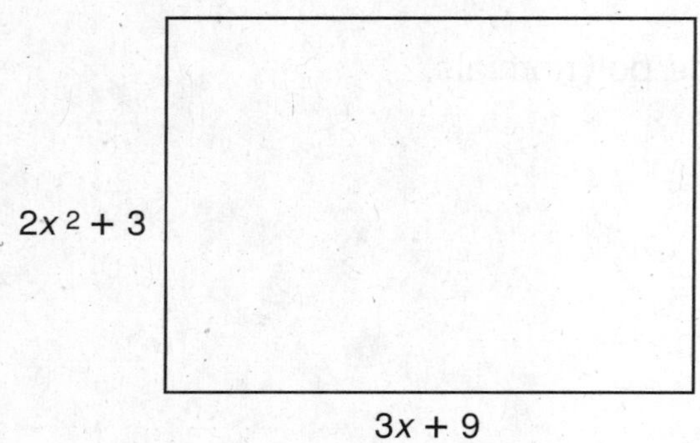


Perimeter: ______________

**4.**

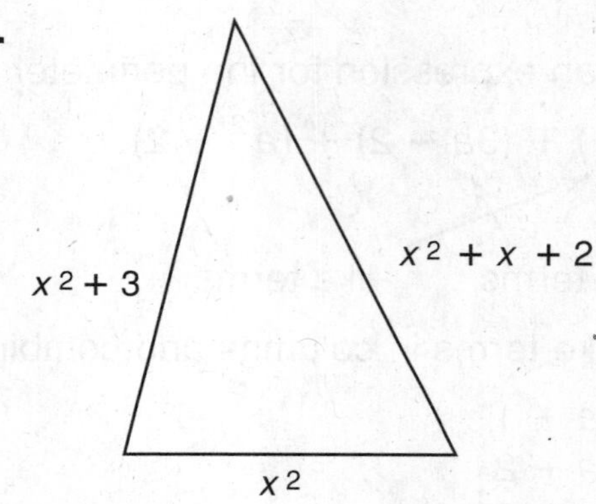


Perimeter: ______________

**5.**

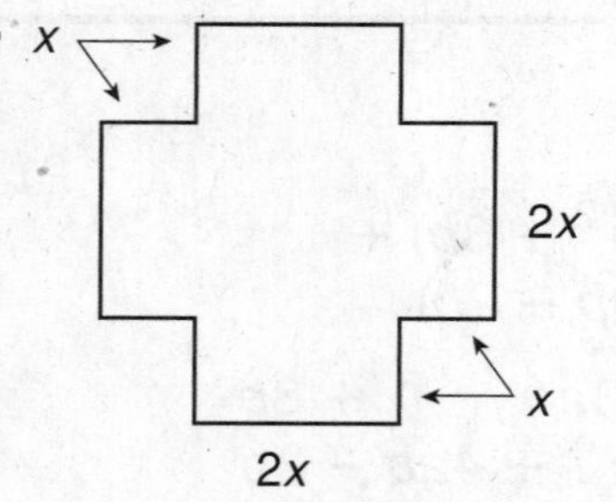


Perimeter: ______________

**6.**

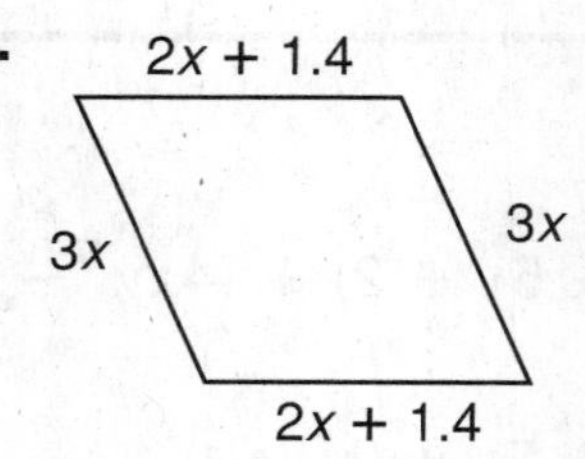


Perimeter: ______________

Name ______________________ Date __________ Class __________

LESSON **14-3**

# Problem Solving

## *Adding Polynomials*

**Write the correct answer.**

1. What is the perimeter of the quadrilateral?

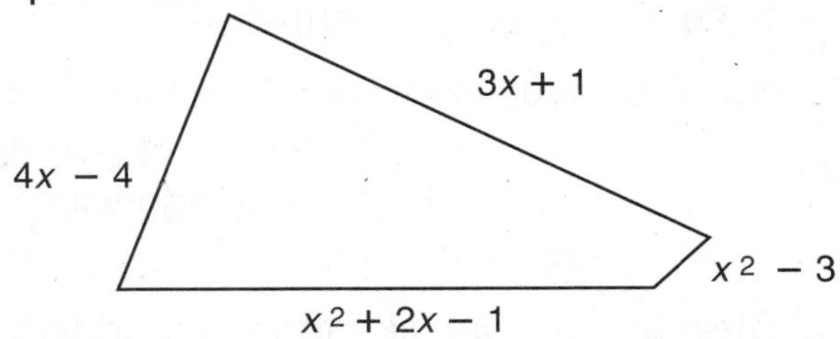


_______________________

2. Jasmine purchased two rugs. One rug covers an area of $x^2 + 8x + 15$ and the other rug covers an area of $x^2 + 3x$. Write and simplify an expression for the combined area of the two rugs.

_______________________

3. Anita's school photo is 12 inches long and 8 inches wide. She will surround the photo with a mat of width $w$. She will surround the mat with a frame that is twice the width of the mat. Find an expression for the perimeter of the framed photo.

_______________________

4. The volume of a right cylinder is given by $\pi r^2 h$. The volume of a right cone is given by $\frac{1}{3}\pi r^2 h$. Write and simplify an expression for the total volume of a right cylinder and right cone combined, if the cylinder and cone have the same radius and height. Use 3.14 for $\pi$.

_______________________

**Choose the letter of the correct answer.**

5. Each side of a square has length $4s - 2$. Which is an expression for the perimeter of the square.

   **A** $8s - 2$
   **B** $16s - 8$
   **C** $8s - 4$
   **D** $16s - 4$

6. The side lengths of a certain triangle can be expressed using the following binomials: $x + 3$, $2x + 2$, and $3x - 2$. Which is an expression for the perimeter of the triangle?

   **F** $2x + 5$
   **G** $2x - 1$
   **H** $3x + 5$
   **J** $6x + 3$

7. What polynomial can be added to $2x^2 + 3x + 1$ to get $2x^2 + 8x$?

   **A** $5x$
   **B** $5x + 1$
   **C** $5x^2 - 1$
   **D** $5x - 1$

8. Which of the following sums is NOT a binomial when simplified?

   **F** $(b^2 + 5b + 1) + (b^2 + 5b + 1)$
   **G** $(b^2 + 5b + 1) + (b^2 + 5b - 1)$
   **H** $(b^2 + 5b + 1) + (b^2 - 5b + 1)$
   **J** $(b^2 + 5b + 1) + (-b^2 + 5b + 1)$

Name ______________________ Date ____________ Class ____________

LESSON 14-3

# Reading Strategies

## *Follow a Procedure*

There are two ways to add polynomials.

| Adding Polynomials Horizontally | Adding Polynomials Vertically |
|---|---|
| **Step 1:** The Associative Property of Addition allows you to regroup the terms.<br>**Step 2:** Identify like terms.<br>**Step 3:** Combine like terms by adding their coefficients.<br>$(7x^3 + 5x^2 - 8) + (6x^2 + 20)$<br>$7x^3 + 5x^2 - 8 + 6x^2 + 20$<br>$7x^3 + \bigcirc{5x^2} - \boxed{8} + \bigcirc{6x^2} + \boxed{20}$<br>$7x^3 + 11x^2 + 12$ | **Step 1:** Identify like terms.<br>**Step 2:** Place like terms in columns. If you rearrange terms, remember to keep the correct sign with each term.<br>**Step 3:** Combine like terms by adding their coefficients.<br>$(2h^2 + 9h^3 + 7) + (11h - 4h^3 - 3h^2)$<br>$9h^3 + 2h^2 \qquad + 7$<br>$-4h^3 - 3h^2 + 11h$<br>$5h^3 - h^2 + 11h + 7$ |

**Use the information in the chart to answer the following questions.**

1. What property allows you to regroup terms in an addition problem?

_______________________________________________

2. How do you combine like terms?

_______________________________________________

3. What happens to a term that has no like terms with which to combine it?

_______________________________________________

4. What are the first steps when adding polynomials vertically?

_______________________________________________

5. How would you rearrange $7 - 4x^2 - 6x + 5x^3$ so that you can vertically add it to $10x^3 + 3x^2 - 9x + 1$?

_______________________________________________

6. When placing like terms in columns to add polynomials vertically, when do you skip spaces?

_______________________________________________

_______________________________________________

Name ______________________ Date __________ Class __________

LESSON 14-3

# Puzzles, Twisters & Teasers

## *It's Amazing!*

**Make your way through the maze by finding the sum of each set of polynomials. You may move up, down, left, right or diagonally, but you may not enter a square more than once.**

**Sum**

1. $(5x^2 - 4x) + (3x^2 + 7x - 2)$ ______________

2. $(-6x^2 + x - 8) + (11x^2 - 5x)$ ______________

3. $(2x^2 + 12x - 14) + (-8x^2 - 15x) + (7x + 20)$ ______________

4. $\begin{array}{r} 10x^2 - 4x + 17 \\ + \ 3x^2 + 9x - 20 \\ \hline \end{array}$ ______________

5. $\begin{array}{r} 6x^3 - 3x^2 + 7x + 4 \\ + \ 5x^3 + 11x^2 \quad\quad - 10 \\ \hline \end{array}$ ______________

| | | | | | | | | | |
|---|---|---|---|---|---|---|---|---|---|
| **$8x^2$** | – | $3x$ | $5x^2$ | + | $4x^2$ | + | 9 | – | + |
| + | $2x$ | 2 | $17x^2$ | – | $6x$ | + | $3x^2$ | 5 | + |
| $7x$ | $3x$ | – | + | $12x^2$ | $4x$ | 13 | + | $8x$ | $6x$ |
| + | + | $4x^2$ | 18 | 12 | – | 6 | $11x^3$ | + | $11x^3$ |
| $13x$ | 2 | – | $-6x^2$ | 8 | 9 | 3 | – | 7 | $4x^2$ |
| $-5x^2$ | $4x$ | + | $6x$ | $13x^2$ | – | $5x$ | – | $6x^2$ | $-8x^2$ |
| $8x^2$ | $4x^2$ | + | 6 | + | 13 | $9x$ | $6x^2$ | $8x^2$ | + |
| – | – | 3 | + | $5x$ | $13x$ | $14x$ | + | $7x^2$ | $7x$ |
| $-2x^2$ | $11x^2$ | + | + | – | 37 | $11x^3$ | – | $5x^2$ | – |
| 8 | $5x$ | $8x$ | $8x^2$ | + | 3 | $11x^2$ | $14x^2$ | + | 6 |

Name ______________________ Date ____________ Class ____________

LESSON 14-4

# Practice A

## *Subtracting Polynomials*

**Find the opposite of each polynomial.**

**1.** $xy^2$

______________

**2.** $-8p^4q^3$

______________

**3.** $5 - 12a$

______________

**4.** $4k^3j + 3k$

______________

**5.** $2u^2 - 6u + 9$

______________

**6.** $-d^4e^3 - 2d^3e^4 - 8de$

______________

**Subtract.**

**7.** $(b^2 + 9) - (3b^2 - 5)$

______________

**8.** $(2x^2 - 3x) - (x^2 + 7)$

______________

**9.** $(-2x^2 + 8x + 9) - (9x^2 + 6x - 2)$

______________

**10.** $(5y^3 + 8y^2 + y) - (9y^3 + 7y^2 - 2)$

______________

**11.** $(5q^5 + q^4 + q^3 + 2q^2 - 5q + 3) - (3 - 2q^3 + q^2 - 10q)$

______________

**12.** $-2f - (f^4 + 3f^3g + 2f^2g - 3fg - 2f - 10)$

______________

**13.** $(nm^2 - 2n^3 + 4n^2m^2 - nm + 15) - (-10nm + 4n^2 - 5nm^2 + 6n^2m^2 - 2n^3)$

______________

**14.** Suppose the number of boxes (in millions) of crayons manufactured annually by the Great Crayon Company is shown by the expression $9x^2 - 8x + 7$. If the rest of the crayon-making companies manufacture $6x^2 - 3x + 4$ crayons, how many more crayons does The Great Crayon Company manufacture each year than the rest of the crayon-making companies?

______________

Name ______________________ Date ____________ Class ____________

LESSON 14-4

# Practice B

## *Subtracting Polynomials*

**Find the opposite of each polynomial.**

**1.** $18xy^3$

**2.** $-9a + 4$

**3.** $6d^2 - 2d - 8$

________________ ________________ ________________

**Subtract.**

**4.** $(4n^3 - 4n + 4n^2) - (6n + 3n^2 - 8)$

**5.** $(-2h^4 + 3h - 4) - (2h - 3h^4 + 2)$

________________________ ________________________

**6.** $(6m + 2m^2 - 7) - (-6m^2 - m - 7)$

**7.** $(17x^2 - x + 3) - (14x^2 + 3x + 5)$

________________________ ________________________

**8.** $w + 7 - (3w^4 + 5w^3 - 7w^2 + 2w - 10)$

________________________________________________

**9.** $(9r^3s - 3rs + 4rs^3 + 5r^2s^2) - (2rs^2 - 2r^2s^2 + 6rs + 7r^3s - 9)$

________________________________________________

**10.** $(3qr^2 - 2 + 14q^2r^2 - 9qr) - (-10qr + 11 - 5qr^2 + 6q^2r^2)$

________________________________________________

**11.** The volume of a rectangular prism, in cubic meters, is given by the expression $x^3 + 7x^2 + 14x + 8$. The volume of a smaller rectangular prism is given by the expression $x^3 + 5x^2 + 6x$. How much greater is the volume of the larger rectangular prism?

________________________________________________

**12.** Sarah has a table with an area, in square inches, given by the expression of $y^2 + 30y + 200$. She has a tablecloth with an area, in square inches, given by the expression of $y^2 + 18y + 80$. She wants the tablecloth to cover the top of the table. What expression represents the number of square inches of additional fabric she needs to cover the top of the table?

________________________________________________

Name ______________________ Date __________ Class __________

LESSON 14-4

# Practice C
## *Subtracting Polynomials*

**Find the opposite of each polynomial.**

**1.** $\frac{1}{8}c^5d^4e$ **2.** $-1.9f + 4g - 2.8h^4$ **3.** $mn^2 + mn - m^2n$

**Subtract.**

**4.** $\left(k^3 - \frac{1}{4}k + \frac{3}{4}k^2\right) - \left(\frac{1}{2}k + \frac{3}{4}k^2 - 6\right)$

**5.** $(100m^2 - 25m^3 + 35) - (94m^2 + 30m + 5m^3 - 25)$

**6.** $(17n^3p - 12np + 4np^3 + 19) - (12np^3 + 6np + 7n^3p - 9)$

**7.** $13q^5r^5 - \left(\frac{1}{25}q^3 + 2r^6 - 10\right) - 11q^5r^5$

**8.** $(3st^2 - 2s^3 + 14s^2t^2) - (+\,4s^2 - 5st^2 + 6s^2t^2) - (-3s^3 + 14s^2)$

**9.** The area of a square, in square yards, is given by the expression $4u^4 + 8u^3 + 12u^2 + 8u + 4$. The area of a smaller square is given by the expression $4u^4 + 4u^3 + 5u^2 + 2u + 1$. How much greater is the area of the larger square?

**10.** The volume of a rectangular prism, in cubic meters, is given by the expression $c^3 + 4c^2 + 3c$. The volume of a smaller rectangular prism is given by the expression $c^3 - c^2 - 2c$. How much greater is the volume of the larger rectangular prism?

Name ______________________ Date __________ Class __________

LESSON 14-4

# Reteach

## *Subtracting Polynomials*

When subtracting polynomials, you can distribute a factor of −1.

Subtract. $(5x^2 + 7x + 3) - (4x^2 + 3x - 5)$.

Rewrite the expression. $(5x^2 + 7x + 3) + (-1)(4x^2 + 3x - 5)$.

Apply the Distributive Property.

$$\mathbf{-1}(4x^2 + 3x - 5) = (\mathbf{-1} \cdot 4x^2) + (\mathbf{-1} \cdot 3x) + (\mathbf{-1} \cdot -5) = -4x^2 - 3x + 5$$

Distributing the −1 changes the sign of each term.
$(5x^2 + 7x + 3) + (-4x^2 - 3x + 5)$

Use the Associative Property to remove parentheses and combine like terms.

$5x^2 + 7x + 3 - 4x^2 - 3x + 5 = x^2 + 4x + 8$

**Subtract.**

**1.** $(3b^3 + 4b^2 + 6) - (b^3 - 5b - 3)$

| | |
|---|---|
| $3b^3 + 4b^2 + 6 + -1(b^3 - 5b - 3)$ | Rewrite the expression. |
| $3b^3 + 4b^2 + 6 + (-b^3 + 5b + 3)$ | Apply the Distributive Property. |
| $3b^3 + 4b^2 + 6 - b^3 + 5b + 3$ | Remove the parentheses. |

________________________________________

**2.** $(3m^2n^2 - 4m^2n + m^2) - (m^2n^2 + 5m^2n - 5)$

________________________________________

**3.** $(2x^3y^2 + x^2y - 4) - (x^2y - 8x + 3)$

________________________________________

**4.** $(6y^2 + 3xy - 9x^2) - (-4y^2 + 8xy + x^2)$

________________________________________

Name ______________________ Date ____________ Class ____________

LESSON 14-4

# Challenge

## Grade a Polynomial Quiz

**Check the student's quiz shown below. If an answer is wrong, write the correct answer.**

Name: Poly Nomial Date: October 5, 2003

**DIRECTIONS: Find each sum. Show all your work.**

**1.** $(2 - 3x + x^2) + (-5 + 7x - 3x^2 + x^3)$

$$\begin{array}{r} x^2 - 3x + 2 \\ + x^3 - 3x^2 + 7x - 5 \\ \hline x^3 - 2x^2 + 10x - 3 \end{array}$$

______________________

**2.** $(-5b^3 + 6b^2 - 1) + (4b^3 + 3b^2 + 2)$

$$\begin{array}{r} -5b^3 + 6b^2 - 1 \\ + 4b^3 + 3b^2 + 2 \\ \hline b^3 + 9b^2 + 1 \end{array}$$

______________________

**3.** $(3m^3 - 4m^2 - 7 + m) + (7m^2 - 4m + 3)$

$$\begin{array}{r} 3m^3 - 4m^2 + m - 7 \\ + \quad 7m^2 - 4m + 3 \\ \hline 3m^3 + 3m^2 - 3m - 4 \end{array}$$

______________________

**4.** $(-8t + 6t^3 - 1 + 4t^2) + (t^3 - 6t^2 + t - 1)$

$$\begin{array}{r} 6t^3 + 4t^2 - 8t - 1 \\ + \quad t^3 - 6t^2 + t - 1 \\ \hline 7t^3 - 2t^2 + 7t - 2 \end{array}$$

______________________

**DIRECTIONS:** Find each difference. Show all your work.

**5.** $(7 + p - 5p^3 + 2p^2) - (-5p^3 + 3p - 7)$

$$\begin{array}{r} -5p^3 + 2p^2 + p + 7 \\ -5p^3 \quad + 3p - 7 \\ \hline -10p^3 + 2p^2 + 4p \end{array}$$

______________________

**6.** $(-a^3 + 3a^2 - 4 + 2a) - (3a^3 + a^2 + 2)$

$$\begin{array}{r} -a^3 + 3a^2 + 2a - 4 \\ - 3a^3 - a^2 \quad - 2 \\ \hline -4a^3 + 2a^2 + 2a - 6 \end{array}$$

______________________

**7.** $(-4h^4 + h^2 - 9h + h^3) - (4h^2 - 3h^4 - 9h)$

$$\begin{array}{r} 4h^4 + h^3 + h^2 - 9h \\ - 3h^4 \quad + 4h^2 - 9h \\ \hline h^4 + h^3 + 5h^2 - 18h \end{array}$$

______________________

**8.** $(-8k + 3k^3 - 6 + k^2) - (5k^3 - k^2 + k^4 - 8)$

$$\begin{array}{r} 3k^3 + k^2 - 8k - 6 \\ - 5k^3 + k^2 - k - 8 \\ \hline -3k^3 + 2k^2 - 9k - 14 \end{array}$$

______________________

Name ______________________ Date __________ Class __________

LESSON **14-4**

# Problem Solving

## *Subtracting Polynomials*

**Write the correct answer.**

1. Molly made a frame for a painting. She cut a rectangle with an area of $x^2 + 3x$ square inches from a piece of wood that had an area of $2x^2 + 9x + 10$ square inches. Write an expression for the area of the remaining frame.

_______________________

2. The volume of a rectangular prism, in cubic inches, is given by the expression $2t^3 + 7t^2 + 3t$. The volume of a smaller rectangular prism is given by the expression $t^3 + 2t^2 + t$. How much greater is the volume of the larger rectangular prism?

_______________________

3. The area of a square piece of cardboard is $4y^2 - 16y + 16$ square feet. A piece of the cardboard with an area of $2y^2 + 2y - 12$ square feet is cut out. Write an expression to show the area of the cardboard that is left.

_______________________

4. A container is filled with $3a^3 + 10a^2 - 8a$ gallons of water. Then $2a^3 - 3a^2 - 3a + 2$ gallons of water are poured out. How much water is left in the container?

_______________________

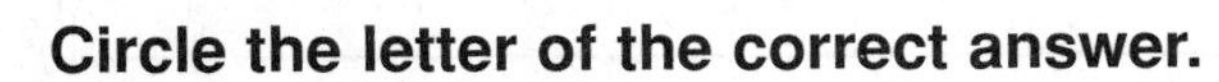

**Circle the letter of the correct answer.**

5. The perimeter of a rectangle is $4x^2 + 2x - 2$ meters. Its length is $x^2 + x - 2$ meters. What is the width of the rectangle?

   **A** $3x^2 + x + 2$ meters

   **B** $2x^2 + 2$ meters

   **C** $x^2 + 1$ meters

   **D** $\frac{3}{2}x - \frac{1}{2}x + 1$ meters

6. On a map, points A, B, and C lie in a straight line. Point A is $x^2 + 2xy + 5y$ miles from Point B. Point C is $3x^2 - 5xy + 2y$ miles from Point A. How far is Point B from Point C?

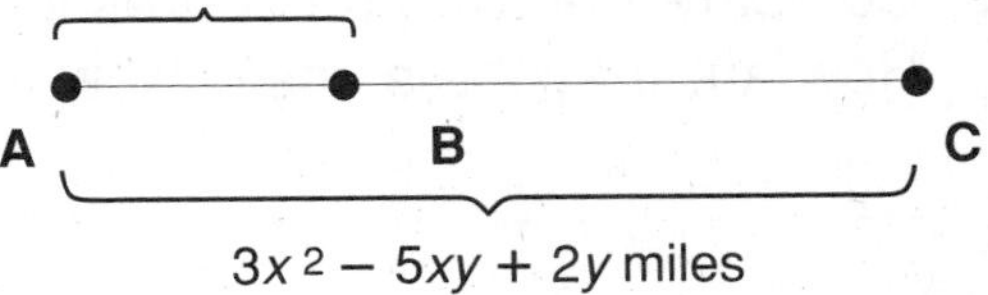


   **F** $-2^2 + 7 + 3y$ miles

   **G** $4x^2 - 3xy + 7y$ miles

   **H** $-4x^2 + 3xy - 7y$ miles

   **J** $2x^2 - 7xy - 3y$ miles

Name ______________________ Date ____________ Class ____________

CHAPTER **14-4**

# Reading Strategies

## *Compare and Contrast*

You can subtract polynomials horizontally or vertically.

| Subtracting Polynomials Horizontally | Subtracting Polynomials Vertically |
|---|---|
| **Step 1:** Rewrite the problem as a sum of the first polynomial and the opposite of the second polynomial.<br>**Step 2:** The Associative Property allows you to regroup the terms.<br>**Step 3:** Identify like terms.<br>**Step 4:** Combine like terms by adding their coefficients.<br>$\mathbf{(2x^3 + 9x^2 + 18) - (5x^2 + 20)}$<br>$(2x^3 + 9x^2 + 18) + (-5x^2 - 20)$<br>$2x^3 + 9x^2 + 18 - 5x^2 - 20$<br>$2x^3 + \bigcirc\!\!\!\!9x^2 + \boxed{18} - \bigcirc\!\!\!\!5x^2 - \boxed{20}$<br>$2x^3 + 4x^2 - 2$ | **Step 1:** Rewrite the problem as a sum of the first polynomial and the opposite of the second polynomial.<br>**Step 2:** Identify like terms.<br>**Step 3:** Place like terms in columns. If you rearrange terms, remember to keep the correct sign with each term.<br>**Step 4:** Combine like terms by adding their coefficients.<br>$\mathbf{(2h^2 + 9h^3 + 7) - (11h - 4h^3 - 3h^2)}$<br>$(2h^2 + 9h^3 + 7) + (-11h + 4h^3 + 3h^2)$<br>$9h^3 + 2h^2 \qquad\quad + 7$<br>$4h^3 + 3h^2 - 11h$<br>$13h^3 + 5h^2 - 11h + 7$ |

**Use the information in the chart to answer the following questions.**

**1.** What is the first step when subtracting polynomials? ______________________

______________________________________________

**2.** How do you find the opposite of a polynomial?

______________________________________________

**3.** Compare the two procedures for subtracting polynomials. How are they the same? How are they different? ______________________

______________________________________________

______________________________________________

______________________________________________

**4.** Compare the process of subtracting polynomials to that of adding polynomials. How are they the same? How are they different?

______________________________________________

______________________________________________

______________________________________________

Name ______________________ Date __________ Class __________

# Puzzles, Twisters & Teasers

## *Digital Displays!*

**How many home runs did Hank Aaron hit in his baseball career?** ________

**To discover the answer, find the missing terms in each of the solutions to the following problems. Then, shade in the sections of the digital display that contain the missing terms. Read the answer from the display.**

1. $(7x^2 + 4x - 5) - (9x + 3) = 7x^2 - 5x -$ ______
2. $(4x^2 - 10x + 17) - (8x + 7) = 4x^2 -$ ______ $+$ ______
3. $(-4x^2 + 6) - (-5x^2 + 3x - 1) =$ ______ $- 3x +$ ______
4. $(3x^3 + 14x^2 + x - 7) - (4x^2 + 6x - 13) =$ ______ $+ 10x^2 - 5x +$ ______
5. $(6x^2 + 11) - (x^2 + 4) - (2x^2 - 6) = 3x^2 +$ ______
6. $(-2x^3 + 7x - 4) - (-6x^3 - 3x^2 + 2x - 5) =$ ______ $+$ ______ $+ 5x +$ ______
7. $(25x^3 + 3) - (6x^2 + 11x) = 25x^3 -$ ______ $-$ ______ $+ 3$

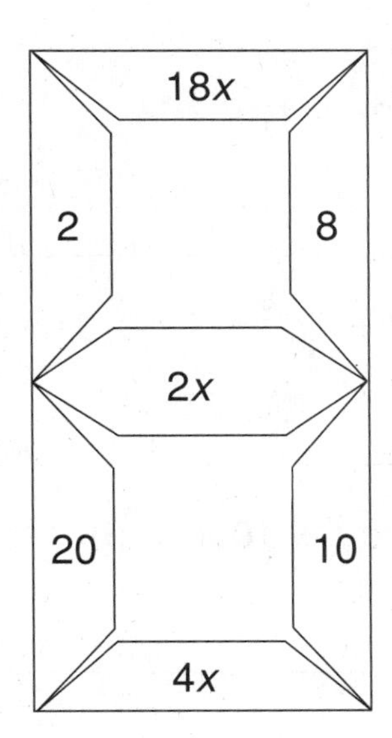


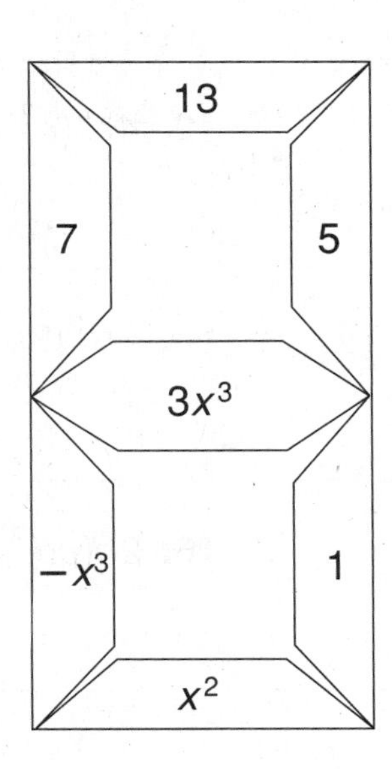


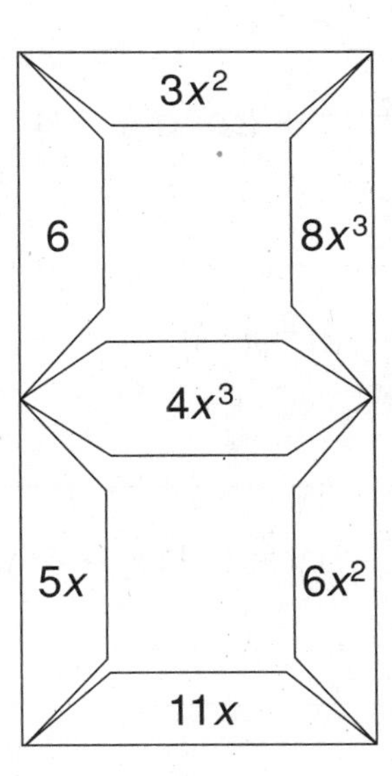

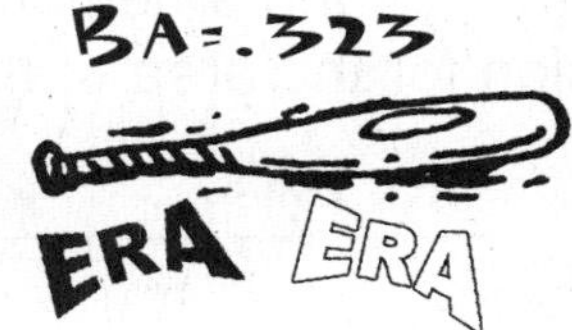

Name ______________________ Date __________ Class __________

LESSON 14-5

# Practice A

## Multiplying Polynomials by Monomials

**Multiply.**

**1.** $(x^2)(-y^3)$

**2.** $(-9z^4)(-z^2)$

**3.** $(-3a^3b^2)(-ab^2)$

**4.** $(-3hi^2)(3h^2i^2)$

**5.** $2(q^2 - 8)$

**6.** $-x(4x^4 - 12)$

**7.** $5y(3y^2 + 2y)$

**8.** $6z\left(-\frac{1}{3}a^5 + 2a\right)$

**9.** $-2jk(4jk + 2j - 2k + 10)$

**10.** $6mn(-m^2 - n^2 + 2mn)$

**11.** $-pq^2(p^2q^2 + 2p^2q + 11)$

**12.** $4r^4(r^2 - 2r + 1)$

**13.** $4s(-s^2 - 2t^2 + 3)$

**14.** $7u^2(3u^2v + 7u^2 - 2u + 1)$

**15.** $xy^2(3x^2 - xy^2 + 11x - 6y)$

**16.** $2d(cd^2 - c^2 + 10cd - 9)$

**17.** A rectangle has a base of length $x^2$ and a height of $3xy^2 - 2x + 1$. Write and simplify an expression for the area of the rectangle.

Name ______________________ Date __________ Class __________

LESSON 14-5

# Practice B

## *Multiplying Polynomials by Monomials*

**Multiply.**

**1.** $(x^2)(-3x^2y^3)$

**2.** $(-9pr^4)(p^2r^2)$

**3.** $(2st^9)(-st^2)$

**4.** $(3efg^2)(-3e^2f^2g)$

**5.** $2q(4q^2 - 2)$

**6.** $-x(x^2 + 2)$

**7.** $5m(-3m^2 + 2m)$

**8.** $6x(-x^5 + 2x^3 + x)$

**9.** $-4st(st - 12t - 2s)$

**10.** $-9ab(a^2 + 2ab - b^2)$

**11.** $-7v^2w^2(vw^2 + 2vw + 1)$

**12.** $8p^4(p^2 - 8p + 17)$

**13.** $4x(-x^2 - 2xy + 3)$

**14.** $7x^2(3x^2y + 7x^2 - 2x)$

**15.** $-4t^3r^2(3t^2r - t^5r - 6t^2r^2)$

**16.** $h^2k(2hk^2 - hk + 7k)$

**17.** A triangle has a base of $4x^2$ and a height of $6x + 3$. Write and simplify an expression for the area of the triangle.

Name ______________________ Date ____________ Class ____________

LESSON 14-5

# Practice C
## *Multiplying Polynomials by Monomials*

**Multiply.**

**1.** $(2g^3)(-7g^2)$

______________________

**2.** $(-9q^2r^4)(q^2r^3)$

______________________

**3.** $\left(\frac{1}{3}s^2t^2\right)(3st)$

______________________

**4.** $\left(\frac{3}{2}g^2\right)\left(-\frac{8}{3}f^2g\right)$

______________________

**5.** $2q(4q^2 - 2q)$

______________________

**6.** $-x^6(x^2 - x + 2)$

______________________

**7.** $\frac{1}{5}m^2n\left(5mn^2 - \frac{3}{5}mn\right)$

______________________

**8.** $x^6(-x^5 + 2x^3 + 7)$

______________________

**9.** $0.2bc(bc - 0.25c)$

______________________

**10.** $-\frac{2}{3}ef\left(e^2 + \frac{1}{2}ef - \frac{3}{5}f^2\right)$

______________________

**11.** $-7v^2w^2(vw^2 + 2vw - 1)$

______________________

**12.** $4x(-x^2 - 2y^2 + 2xy + x + 3)$

______________________

**13.** $2y^2(-y^2z^2 - 2yz^2 + 4y^2 + 2z^2)$

______________________

**14.** $-4h^3k^2(3h^4k - h^2k + 2h^3 + k^6)$

______________________

**15.** A square has a side of length $2y^2$. Write and simplify an expression for the area of the square. Then find the area of the square if $y = 3$.

______________________

**16.** A rectangle has a base of $3m^2 + m - 2$ and height of $4m^2$. Write and simplify an expression for the area of the rectangle. Then find the area of the rectangle if $m = 2$.

______________________

______________________

Name ______________________ Date __________ Class __________

LESSON 14-5

# Reteach

## *Multiplying Polynomials by Monomials*

To multiply a monomial by a monomial, follow the steps used in the example below.

$(7x^2y^3)(3xy^4)$

**1.** Multiply the coefficients.

$(7)(3) = 21$

**2.** Multiply the variables.

To multiply two powers with the same base, you keep the base and **add** the exponents.

$(x^2)(x) = (x^2)(x^1) = x^3 \qquad (y^3)(y^4) = y^7$

**3.** Write the monomial product.

$21x^3y^7$

Remember: If a variable has no exponent, the exponent is 1.
$x = x^1$

**Multiply.**

**1.** $(3x^2)(4x^3y^2)$ ______________

**2.** $(6a^3b)(2a^3b^4)$ ______________

**3.** $(2m^4n^2)(-5m^2n^2)$ ______________

To multiply a polynomial by a monomial, multiply each term of the polynomial by the monomial.

$$\begin{array}{r} 4a^2 + 2ab + 6b^2 \\ \times \qquad\qquad\quad 3a^3 \\ \hline 12a^5 + 6a^4b + 18a^3b^2 \end{array}$$

**Multiply.**

**4.** $\begin{array}{r} 3r^2s^3 - 2r^2 + 10 \\ \times \qquad\qquad 2s \end{array}$

______________

______________

**5.** $\begin{array}{r} 5x^5 + x^2 - 3x \\ \times \qquad\qquad 4x^3 \end{array}$

______________

______________

**6.** $\begin{array}{r} m^2n - 3mn^2 - 8n^3 \\ \times \qquad\qquad -3mn \end{array}$

______________

______________

Name ______________________ Date ____________ Class ____________

LESSON 14-5

## Challenge

### *X Marks the Surface*

**Write a polynomial for the surface area of each box. Simplify each polynomial. Then use your polynomials to find each box's surface area, if $x = 4$ inches. Remember: For a rectangular prism, $SA = 2lw + 2lh + 2hw$.**

**1.**

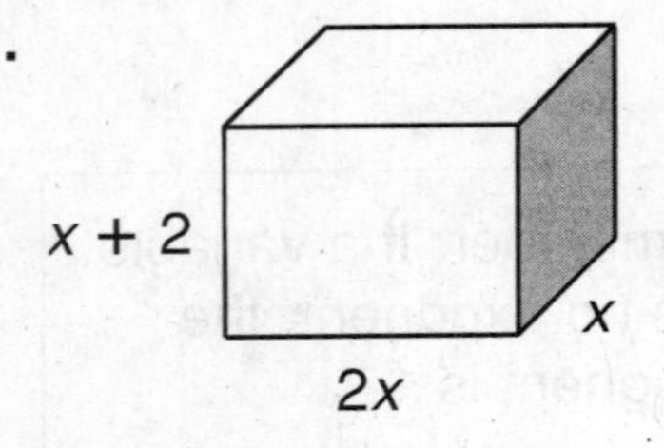


______________________

SA = ______________________

**2.**

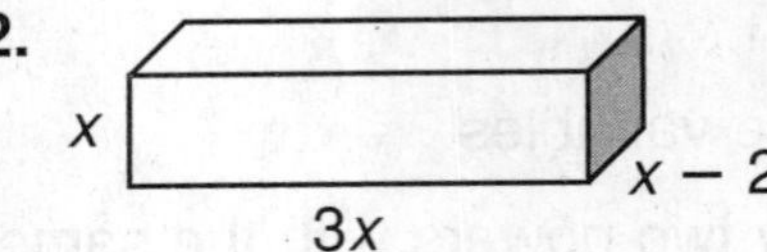


______________________

SA = ______________________

**3.**

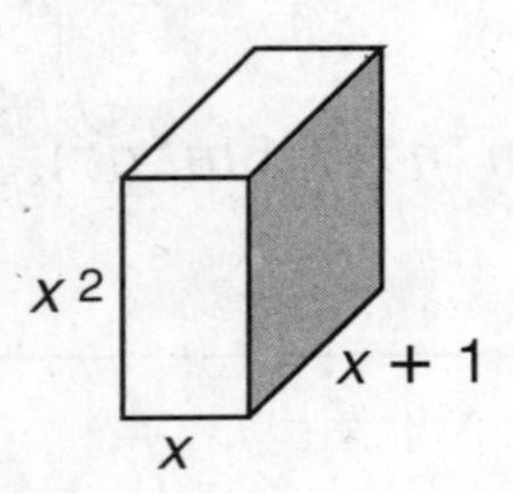


______________________

SA = ______________________

**4.**

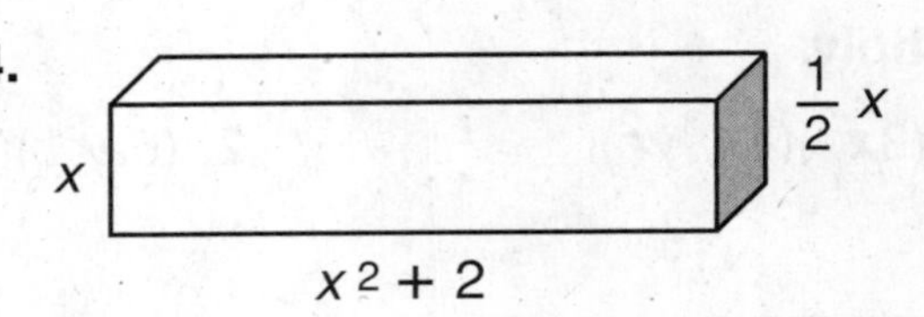


______________________

SA = ______________________

**5.**

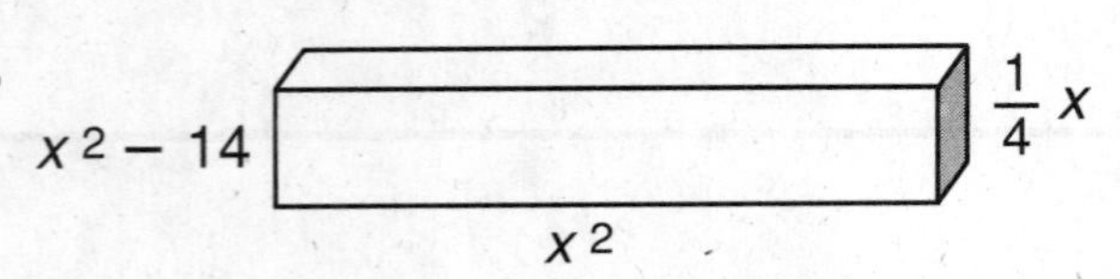


______________________

SA = ______________________

**6.**

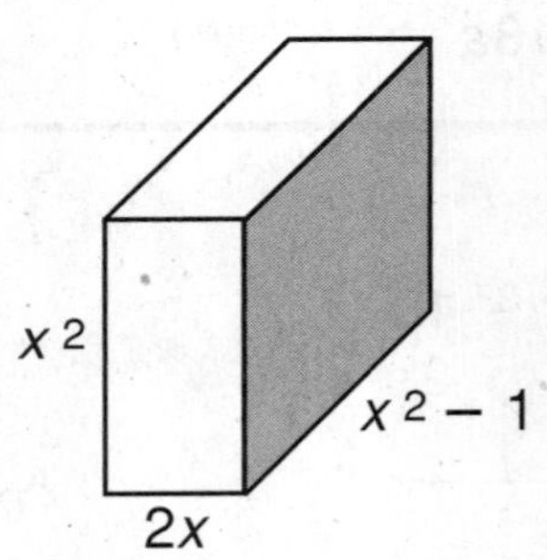


______________________

SA = ______________________

Name ______________________ Date __________ Class __________

LESSON 14-5

# Problem Solving

## Multiplying Polynomials by Monomials

**Write the correct answer.**

**1.** A rectangle has a width of $5n^2$ inches and a length of $3n^2 + 2n + 1$ inches. Write and simplify an expression for the area of the rectangle. Then find the area of the rectangle if $n = 2$ inches.

_______________________________

_______________________________

**2.** The area of a parallelogram is found by multiplying the base and the height. Write and simplify an expression for the area of the parallelogram below.

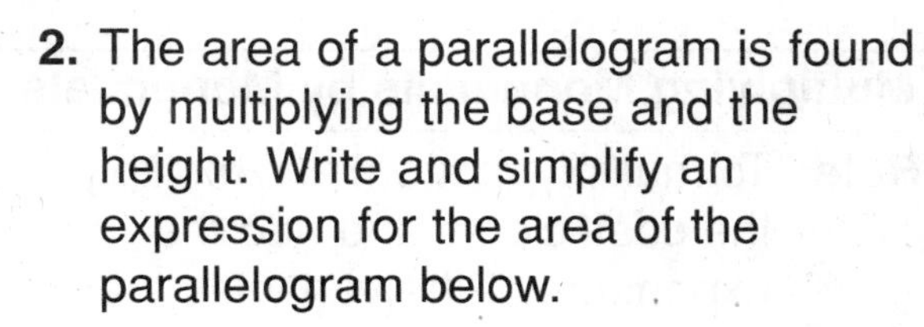

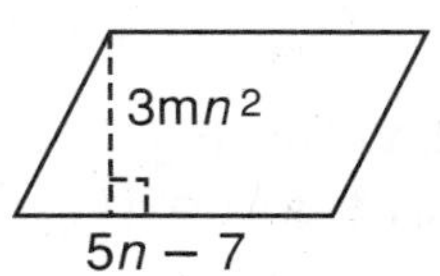


_______________________________

_______________________________

**3.** A parallelogram has a base of $2x^2$ inches and a height of $x^2 + 2x - 1$ inches. Write an expression for the area of the parallelogram. What is the area of the parallelogram if $x = 2$ inches?

_______________________________

_______________________________

**4.** A rectangle has a length of $x^2 + 2x - 1$ meters and a width of $x^2$ meters. Write an expression for the area of the rectangle. What is the area of the rectangle if $x = 3$ meters?

_______________________________

_______________________________

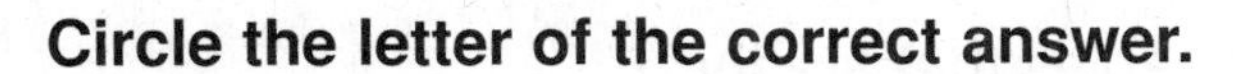

**Circle the letter of the correct answer.**

**5.** A rectangle has a width of $3x$ feet. Its length is $2x + \frac{1}{6}$ feet. Which expression shows the area of the rectangle?

**A** $5x + \frac{1}{6}$

**B** $6x^2 + \frac{1}{2}x^2$

**C** $6x^2 + \frac{1}{2}$

**D** $6x^2 + \frac{1}{2}x$

**6.** Which expression shows the area of the shaded region of the drawing?

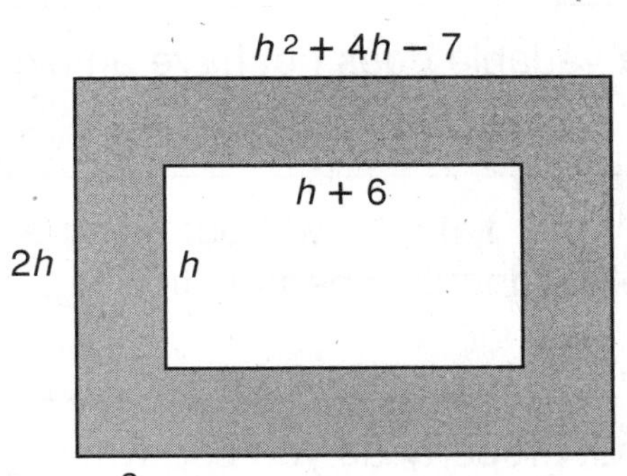


**F** $2h^3 + 8h - 14h$

**G** $2h^3 + 9h^2 - 8h$

**H** $2h^3 + 7h^2 - 20h$

**J** $2h^3 + 7h^2 - 8h$

Name ______________________ Date ____________ Class ____________

LESSON **14-5**

# Reading Strategies

## *Draw Conclusions*

Follow these rules to multiply polynomials.

| Multiplying Monomials by Monomials | Multiplying Polynomials by Monomials |
|---|---|
| **Rule:** To multiply monomials, multiply the coefficients and add the exponents of the variables that are the same.<br><br>Remember, if a variable does not have an exponent, the exponent is 1.<br><br>**Example:** $(12x^6y^2)(4xy^3)$<br>$(12x^6y^2)(4xy^3)$<br>$12 \cdot 4 \cdot x^{6+1}y^{2+3}$<br>$48x^7y^5$ | **Rule:** To multiply a polynomial by a monomial, use the Distributive Property. Multiply every term of the polynomial by the monomial.<br><br><br>**Example:** $\mathbf{-6ab^5(2a^2 + 4ab - 9b)}$<br>$-6ab^5(2a^2 + 4ab - 9b)$<br>$-12a^3b^5 - 24a^2b^6 + 54ab^6$ |

**Use the information in the chart to answer the following questions.**

1. When multiplying two or more monomials, what do you do to the coefficients of the variables?

_______________________________________________

2. What do you do to the exponents of the variables that are the same?

_______________________________________________

3. If a variable does not have an exponent, what is the exponent?

_______________________________________________

4. Write $y^2$ times $y^3$ without using exponents to show why you add the exponents when multiplying.

_______________________________________________

5. What property do you use when multiplying a polynomial by a monomial?

_______________________________________________

6. How many terms will you get if you multiply $2x^2y$ by $(4xy^3 + 2x - 5)$?

_______________________________________________

Name ______________________ Date ____________ Class ____________

LESSON 14-5

# Puzzles, Twisters & Teasers

## *Tic–Tac–Nomials!*

**Simplify the expressions in each square to determine whether they are equal. If they are equal, place an X in the square. If they are not equal, place an O in the square. Be sure to mark the Tic–Tac–Toe!**

| | | |
|---|---|---|
| $(12x^2y)(2x^3y^4)$ <br> $\overset{?}{=}$ <br> $(6x)(4x^4y)$ | $2a^2b^2(20a + 8ab - 6b^2)$ <br> $\overset{?}{=}$ <br> $4b^2(10a^3 + 4a^3b - 3a^2b^2)$ | $-6m(4 + 8m^2)$ <br> $\overset{?}{=}$ <br> $2m(-12 + 24m^2)$ |
| $-2c(4c^2d) + 10c^3d$ <br> $\overset{?}{=}$ <br> $2c(c^2d)$ | $(3rs^3)(-8r^4s)$ <br> $\overset{?}{=}$ <br> $(-6r^5s^2)(4s^2)$ | $-5x(4x^2y - 6xy + 8y^2)$ <br> $\overset{?}{=}$ <br> $10y(3x^2 - 2x^3 - 4xy)$ |
| $-4gh^2(9g^2 + 6h^2)$ <br> $\overset{?}{=}$ <br> $12g(-2h^4 - 3g^2h^2)$ | $6a(a^2 - 10a) + 50a^2$ <br> $\overset{?}{=}$ <br> $2a^2(3a + 5)$ | $(12xy)(-6x^3y)$ <br> $\overset{?}{=}$ <br> $(-8x^2y^2)(-9x^2)$ |

Name ______________________ Date __________ Class __________

LESSON 14-6

# Practice A
## *Multiplying Binomials*

**Multiply.**

**1.** $(z + 3)(z + 2)$ ____________

**2.** $(a + 1)(a + 2)$ ____________

**3.** $(b + 1)(3b + 1)$ ____________

**4.** $(c + 1)(c - 3)$ ____________

**5.** $(3v - 1)(3v - 3)$ ____________

**6.** $(2r + 1)(2r - 1)$ ____________

**7.** $(u - 5v)(3u - v)$ ____________

**8.** $(-3g + 4)(g - 1)$ ____________

**9.** $(-2r + s)(r + 6s)$ ____________

**10.** Write an expression for the area of a cement walkway of width $k$ ft around an 8 ft by 10 ft jacuzzi.

____________________________________

**11.** Write and simplify an expression for the area of a brick walkway of width $x$ m around a 30 m by 50 m yard.

____________________________________

**12.** A rug is placed in a 15 ft × 25 ft room so that there is an uncovered strip of width $h$ feet all the way around the rug. Write and simplify an expression for the area of the rug.

____________________________________

**Multiply.**

**13.** $(e + 1)^2$ ____________

**14.** $(2a + 1)^2$ ____________

**15.** $(e - 3)^2$ ____________

**16.** $(m - 1)(m + 1)$ ____________

**17.** $(2 + e)^2$ ____________

**18.** $(k - n)^2$ ____________

**19.** $(2w - z)(2w + z)$ ____________

**20.** $(2a + b)(2a - b)$ ____________

Name ______________________ Date __________ Class __________

LESSON 14-6

# Practice B
## *Multiplying Binomials*

**Multiply.**

1. $(z + 1)(z + 2)$ ______

2. $(1 - y)(2 - y)$ ______

3. $(2x + 1)(2x + 4)$ ______

4. $(w + 1)(w - 3)$ ______

5. $(3v + 1)(v - 1)$ ______

6. $(t + 2)(2t - 2)$ ______

7. $(-3g + 4)(2g - 1)$ ______

8. $(3c + d)(c - 2d)$ ______

9. $(2a + b)(a + 2b)$ ______

10. A box is formed from a 1 in. by 18 in. piece of cardboard by cutting a square with side length $m$ inches out of each corner and folding up the sides. Write and simplify an expression for the area of the base of the box.

______

11. A table is placed in a 14 ft × 18 ft room so that there is an equal amount of space of width $s$ feet all the way around the table. Write and simplify an expression for the area of the table.

______

12. A circular swimming pool with a radius of 14 ft is surrounded by a deck with width $y$ feet. Write and simplify an expression for the total area of the pool and the deck. Use $\frac{22}{7}$ for pi.

______

**Multiply.**

13. $(r - 2)^2$ ______

14. $(2 + q)^2$ ______

15. $(p + 4)(p - 4)$ ______

16. $(3n - 3)(3n + 3)$ ______

17. $(a + b)(a - b)$ ______

18. $(4e - f)^2$ ______

19. $(2y + z)^2$ ______

20. $(9p - 2)(-2 + 9p)$ ______

21. $(m - 1)^2$ ______

Name ______________________ Date __________ Class __________

LESSON 14-6

# Practice C
## *Multiplying Binomials*

**Multiply.**

**1.** $(a + 1)(a + 2)$ ____________

**2.** $(1 - x)(2 - x)$ ____________

**3.** $(2c + 1)(2c + 4)$ ____________

**4.** $(2w + 1)(-2w - 3)$ ____________

**5.** $\left(2v + \frac{1}{4}\right)\left(v - \frac{1}{4}\right)$ ____________

**6.** $(3t + 2)(2t - 3)$ ____________

**7.** A circular swimming pool is surrounded by a deck that is $y$ meters wide. The radius of the swimming pool is 21 meters. Write and simplify an expression for the total area of the swimming pool and deck. Use $\frac{22}{7}$ for pi.

____________________________________________

**8.** A rug is placed in a 6 ft × 15 ft hallway so that there is an equal amount of space of width $y$ feet all the way around the rug. Write and simplify an expression for the area of the rug.

____________________________________________

**9.** A brick walkway around a circular jacuzzi is 7 ft wide. The radius of the jacuzzi is $14d$ ft. Write and simplify an expression for the total area of the jacuzzi and the brick walkway around the jacuzzi. Use $\frac{22}{7}$ for pi.

____________________________________________

**Multiply.**

**10.** $(p + 2)^2$ ____________

**11.** $(2 - k)^2$ ____________

**12.** $(d + 14)(d - 14)$ ____________

**13.** $(3c - 3d)(3c + 3d)$ ____________

**14.** $(2a + b)(2a - b)$ ____________

**15.** $\left(\frac{1}{2}c + d\right)\left(\frac{1}{2}c - d\right)$ ____________

**16.** $(9p - 2q)(-2q + 9p)$ ____________

**17.** $(2m - 3b)^2$ ____________

**18.** $(9a + 3b)(9a - 3b)$ ____________

Name ______________________ Date ____________ Class ____________

LESSON 14-6

# Reteach

## Multiplying Binomials

To multiply a binomial by a binomial, multiply each term of the first binomial by each term of the second binomial.

$$(a + b)(c + d) = ac + ad + bc + bd$$

You can remember the product as FOIL: First terms, Outer terms, Inner terms, and Last terms.

$(5x + 3)(3x - 2)$

| | |
|---|---|
| Multiply the **F**irst terms. | $(5x)(3x) = 15x^2$ |
| Multiply the **O**utside terms. | $(5x)(-2) = -10x$ |
| Multiply the **I**nside terms. | $(3)(3x) = 9x$ |
| Multiply the **L**ast terms. | $(3)(-2) = -6$ |
| Add the products. | $15x^2 - 10x + 9x - 6$ |
| Combine like terms. | $15x^2 - x - 6$ |

**You can also multiply binomials vertically.**

Align the binomials.

Multiply each term of one binomial by each term of the other binomial.

Combine like terms.

$$\begin{array}{r} 5x + 3 \\ \times\ 3x - 2 \\ \hline -10x - 6 \\ 15x^2 + 9x \qquad \\ \hline 15x^2 - \ \ x - 6 \end{array}$$

**Multiply.**

1. $(4x + 3)(2x + 5)$ ______________________

2. $(7t - 4)(2t + 3)$ ______________________

3. $(3 + 5b)(2b - 3b^2)$ ______________________

4. $(x - 1)(x + 5)$ ______________________

5. $(6m - 3n)(2m + 3n)$ ______________________

6. $(c + 7)(c + 1)$ ______________________

7. $\begin{array}{r} 6n - 3 \\ \times\ 3n + 3 \\ \hline \end{array}$ ______________________

8. $\begin{array}{r} 2y + 4 \\ \times\ \ y + 6 \\ \hline \end{array}$ ______________________

Name ______________________ Date __________ Class __________

LESSON 14-6

# Challenge
## *Multiplication Tables*

You can use a table to multiply binomials. Label each column with a term from one of the binomials, and label each row with a term from the other binomial of the coefficients. Then multiply the same as in a multiplication table (column × row). Finish by combining like terms in the product.

**Example: $(x + 3)(x - 2)$**

| | **$x$** | **3** |
|---|---|---|
| **$x$** | $x^2$ | $3x$ |
| **−2** | $-2x$ | $-6$ |

$x^2 + 3x - 2x - 6 = \mathbf{x^2 + x - 6}$

**Use the given tables to find each product.**

**1.** $(x - 4)(x - 5)$

| | **$x$** | **−4** |
|---|---|---|
| **$x$** | | |
| **−5** | | |

______________________

**2.** $(3m + 5)^2$

| | **$3m$** | **5** |
|---|---|---|
| **$3m$** | | |
| **5** | | |

______________________

**3.** $(2h + 1)(h - 4)$

| | | |
|---|---|---|
| | | |
| | | |

______________________

**4.** $(7p - q)(p + 6q)$

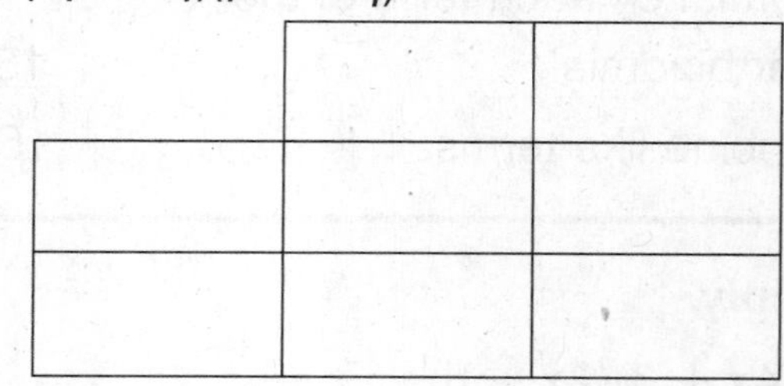

______________________

**5.** $(3a + 4b)^2$

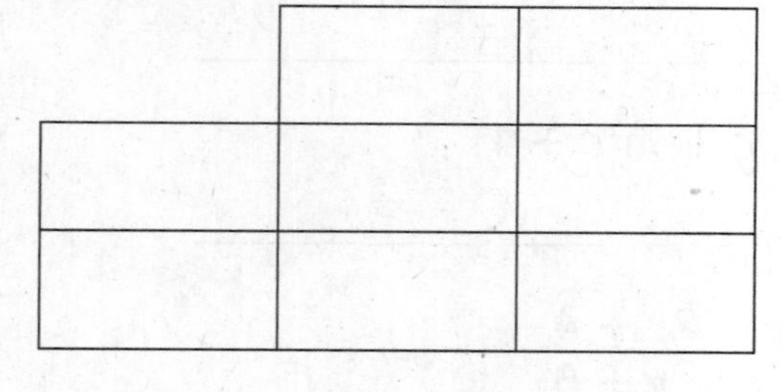

______________________

**6.** $(5w - 10)(2w - v)$

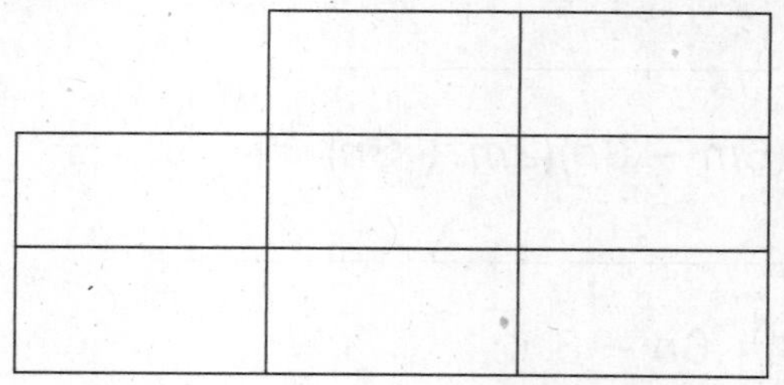

______________________

Name ______________________ Date __________ Class __________

LESSON **14-6**

# Problem Solving

## *Multiplying Binomials*

**Write and simplify an expression for the area of each polygon.**

| | Polygon | Dimensions | Area |
|---|---|---|---|
| **1.** | rectangle | length: $(n + 5)$; width: $(n - 4)$ | |
| **2.** | rectangle | length: $(3y + 3)$; width: $(2y - 1)$ | |
| **3.** | triangle | base: $(2b - 5)$; height: $(b^2 + 2)$ | |
| **4.** | square | side length: $(m + 13)$ | |
| **5.** | square | side length: $(2g - 4)$ | |
| **6.** | circle | radius: $(3c + 2)$ | |

**Choose the letter of the correct answer.**

**7.** A photo is 8 inches by 11 inches. A frame of width $x$ inches is placed around the photo. Which expression shows the total area of the frame and photo?

**A** $x^2 + 19x + 88$

**B** $4x^2 + 38x + 88$

**C** $8x + 38$

**D** $4x + 19$

**8.** Three consecutive odd integers are represented by the expressions, $x$, $(x + 2)$ and $(x + 4)$. Which expression gives the product of the three odd integers?

**F** $x^3 + 8$

**G** $x^3 + 6x^2 + 8x$

**H** $x^3 + 6x^2 + 8$

**J** $x^3 + 2x^2 + 8x$

**9.** A square garden has a side length of $(b - 4)$ yards. Which expression shows the area of the garden?

**A** $2b - 8$

**B** $b^2 + 16$

**C** $b^2 - 8b - 16$

**D** $b^2 - 8b + 16$

**10.** Which expression gives the product of $(3m + 4)$ and $(9m - 2)$?

**F** $27m^2 + 30m - 8$

**G** $27m^2 + 42m - 8$

**H** $27m^2 + 42m + 8$

**J** $27m^2 + 30m + 8$

Name ______________________ Date __________ Class __________

LESSON **14-6**

# Reading Strategies

## *Use Graphic Aids*

This chart can help you remember how to multiply binomials.

| How to Multiply Binomials | Special Products of Binomials |
|---|---|
| **F**IRST, **O**UTER, **I**NNER, **L**AST 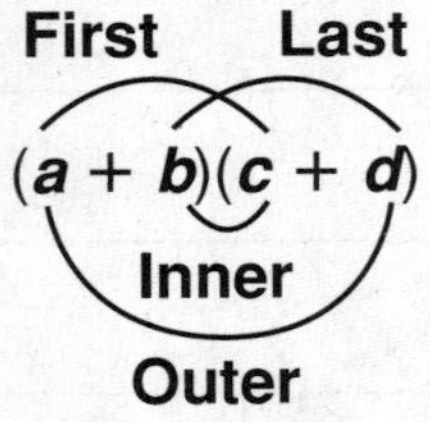<br>**First** **Last**<br>$(a + b)(c + d)$<br>**Inner**<br>**Outer**<br>$ac + ad + bc + bd$<br>The Commutative Property states that it does not matter in which order you write the addends. | $(a + b)^2 = (a + b)(a + b) = a^2 + 2ab + b^2$<br>**Example:** $(x + 4)^2 = x^2 + 2(4)(x) + 4^2 = x^2 + 8x + 16$<br>$(a - b)^2 = (a - b)(a - b) = a^2 - 2ab + b^2$<br>**Example:** $(x - 6)^2 = x^2 - 2(6)(x) + 6^2 = x^2 - 12x + 36$<br>$(a + b)(a - b) = a^2 - b^2$<br>**Example:** $(x + 7)(x - 7) = x^2 - 7^2 = x^2 - 49$ |

**Use the information in the chart to answer the following questions.**

**1.** What do the letters FOIL represent?

______________________________________________

**2.** Does it matter in which order you write the addends? Why or why not??

______________________________________________

**3.** When finding the special product of $(a + b)^2$, how many terms will you always have?

______________________________________________

**4.** When finding the special product of $(a - b)^2$, why is the last term positive?

______________________________________________

**5.** Why do you get only two terms when you multiply binomials in the special form $(a + b)(a - b)$?

______________________________________________

Name ______________________ Date __________ Class __________

LESSON **14-6**

# Puzzles, Twisters & Teasers

## *Back It Up!*

**What do you have when a row of rabbits steps backwards?**

**To find the answer, find each product in Column 1 and match it to the correct expression in Column 2. Then, write the letter above the corresponding exercise number.**

| Column 1 | Column 2 |
|---|---|
| **1.** $(x + 4)^2$ | **A** $x^2 + 4x - 32$ |
| **2.** $(x + 3)(x + 6)$ | **C** $x^2 + 9x + 18$ |
| **3.** $(x + 8)(x - 4)$ | **D** $x^2 - 16$ |
| **4.** $(x + 4)(x - 4)$ | **E** $6x^2 - 2x - 20$ |
| **5.** $(x - 9)(x - 2)$ | **G** $9x^2 - 30x + 25$ |
| **6.** $(x - 16)(x + 2)$ | **H** $x^2 - 14x - 32$ |
| **7.** $(2x - 5)(3x + 4)$ | **I** $6x^2 - 7x - 20$ |
| **8.** $(6x + 10)(x - 2)$ | **L** $x^2 + 8x + 16$ |
| **9.** $(3x - 5)^2$ | **N** $x^2 - 11x + 18$ |
| **10.** $(3x - 5)(3x + 5)$ | **R** $9x^2 - 25$ |

| ___ | ___ | ___ | ___ | ___ | ___ | ___ | ___ | ___ |
|---|---|---|---|---|---|---|---|---|
| 3 | 10 | 8 | 2 | 8 | 4 | 7 | 5 | 9 |

| ___ | ___ | ___ | ___ | ___ | ___ | ___ | ___ |
|---|---|---|---|---|---|---|---|
| 6 | 3 | 10 | 8 | 1 | 7 | 5 | 8 |

## LESSON 14-1 Practice A
### Polynomials

**Determine whether each expression is a monomial.**

1. $-x^3$ — yes
2. $4xy^9$ — yes
3. $\frac{8}{q^3}$ — no
4. $3.5r^{\frac{1}{2}}$ — no
5. $a^{2.2}$ — no
6. $\frac{7}{9}xyz$ — yes

**Classify each expression as a monomial, a binomial, a trinomial, or not a polynomial.**

7. $-8.9x + 6x^5$ — binomial
8. $\frac{1}{18}a^8a^2$ — monomial
9. $x^8 + x + \frac{1}{x}$ — not a polynomial
10. $-7r^4 + s^3$ — not a polynomial
11. $m^{15} - m + \frac{1}{9}$ — trinomial
12. $-7.55r^{75}$ — monomial

**Find the degree of each polynomial.**

13. $\frac{7}{2} - x$ — 1
14. $a + a^2 + a^3$ — 3
15. $7w - 6w + 3w$ — 1
16. $-9p - 9p - 9p^3$ — 3
17. $y^9 + 10y^8 - 7y$ — 9
18. $1{,}055 + \frac{4}{5}k - k^7$ — 7

19. The volume of a box with width $x$, length $2x + 1$, and height $2x - 2$ is given by the trinomial $4x^3 - 2x^2 - 2x$. What is the volume of the box if its width is 5 meters?

440 m$^3$




## LESSON 14-1 Practice B
### Polynomials

**Determine whether each expression is a monomial.**

1. $-135x^5$ — yes
2. $2.4x^3y^{19}$ — yes
3. $\frac{2p^2}{q^3}$ — no
4. $3r^{\frac{1}{2}}$ — no
5. $43a^2b^{6.1}$ — no
6. $\frac{7}{9}x^2yz^5$ — yes

**Classify each expression as a monomial, a binomial, a trinomial, or not a polynomial.**

7. $-8.9xy + \frac{6}{y^5}$ — not a polynomial
8. $\frac{9}{8}ab^8c^2d$ — monomial
9. $x^8 + x + 1$ — trinomial
10. $-7pq^{-2}r^4$ — not a polynomial
11. $5n^{15} - 9n + \frac{1}{3}$ — trinomial
12. $r^8 - 5.5r^{75}$ — binomial

**Find the degree of each polynomial.**

13. $7 - 14x$ — 1
14. $5a + a^2 + \frac{6}{7}a^3$ — 3
15. $7w - 16u + 3v$ — 1
16. $9p - 9q - 9p^3 - 9q^2$ — 3
17. $z^9 + 10y^8 - x$ — 9
18. $100{,}050 + \frac{4}{5}k - k^4$ — 4

19. The volume of a box with height $x$, length $x - 1$, and width $2x + 2$ is given by the binomial $2x^3 - 2x$. What is the volume of the box if its height is 4 feet?

120 ft$^3$

20. The trinomial $-16t^2 + 32t + 32$ describes the height in feet of a ball thrown upward after $t$ seconds. What is the height of the ball $\frac{5}{8}$ seconds after it was thrown?

45.75 feet

/20




## LESSON 14-1 Practice C
### Polynomials

**Determine whether each expression is a monomial.**

1. $-1.35a^{135}b^{12}c^3$ — yes
2. $2.4x^{\frac{3}{2}}y^{19}$ — no
3. $\frac{2p^2q^5}{pq^3}$ — yes

**Classify each expression as a monomial, a binomial, a trinomial, or not a polynomial.**

4. $-0.9x + \frac{6}{y^5} + 1$ — not a polynomial
5. $\frac{8ab^8c^2d}{ad}$ — monomial
6. $-8x^8 + x + 15x^6$ — trinomial
7. $-7h^{-2}k$ — not a polynomial
8. $5m^{15} - 9n^9 - 0.03$ — trinomial
9. $0.6t^6 - 75g^0t^{75}$ — binomial

**Find the degree of each polynomial.**

10. $50 - 1.024s^6$ — 6
11. $5u^2 + u + \frac{6}{5}u^3$ — 3
12. $\frac{7}{1}m - 2.05n + 3p$ — 1
13. $9(p - \frac{2}{3}q - 9p^3)$ — 3
14. $(z^9)^0 + 10y^{89} - x$ — 89
15. $\frac{4}{11}h^4k - k^4$ — 4

16. The area of a rectangle with length $x$ and width $2x - 3$ is given by the binomial $2x^2 - 3x$. What is the area of the rectangle if its length is 5 yards?

35 yd$^2$

17. The height in feet of a ball thrown straight up into the air from $s$ feet off the ground at velocity $v$ after $t$ seconds is given by the polynomial $-16t^2 + vt + s$. Find the height of a ball thrown from a 20 ft platform at 150 ft/s after 6 seconds.

344 feet




## LESSON 14-1 Reteach
### Polynomials

Expressions such as $2x$ and $4y^2$ are called **monomials.** A monomial has only one term. Monomials do not have fractional exponents, negative exponents, variable exponents, roots of variables, or variables in a denominator.

/19

**Determine whether each expression is a monomial.**

1. $3x - 5$ — no
2. $-9a^4$ — yes
3. $21m^{0.5}$ — no
4. $7m^3n^2$ — yes

A monomial or a sum or difference of monomials is called a **polynomial.** Polynomials can be classified by the number of terms. A monomial has 1 term, a **binomial** has 2 terms, and a **trinomial** has 3 terms.

**Classify each expression as a monomial, a binomial, a trinomial, or not a polynomial.**

5. $7y + 3x^2 + 5$ — trinomial
6. $6y + \sqrt{x}$ — not a polynomial
7. $m^2n$ — monomial
8. $-6a + 2b^4$ — binomial

The degree of a polynomial is the degree of the term with the greatest degree. The **degree** of a term is the greatest value of a variable's exponent.

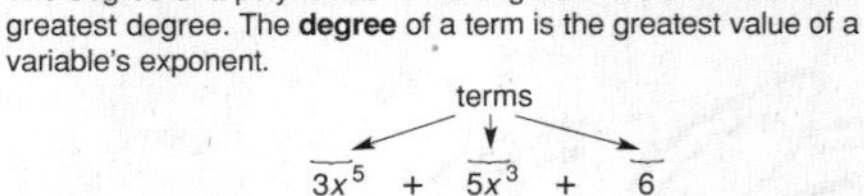


**5th degree** 3rd degree 0 degree

The above polynomial is a 5th degree trinomial.

**Find the degree of each polynomial.**

9. $5x + 3x^3 + 2x^2$ — 3
10. $-3m^4 + m^2 + 2$ — 4
11. $4y + 2y^3 + y^5$ — 5
12. $7a^2 + 8a$ — 2

/12

LESSON 14-1

## Challenge

### Polynomial Means "Many Names"

How many different ways can you name the expression $3x^3 + 4x + 7$? You can call it a polynomial because it is a sum of monomials, or a trinomial because it is a polynomial with 3 terms. You can also call it a polynomial of third degree because the term with the greatest degree has degree 3. But did you know that some polynomials have even more names?

**For each polynomial described in the chart below, write an example of a monomial, a binomial, and a trinomial to match its special degree name.** Possible answers are given.

| Degree | Names | Monomial | Binomial | Trinomial |
|---|---|---|---|---|
| 1 | Linear or Monic | $x$ | $x + y$ | $x + y + 1$ |
| 2 | Quadratic | $x^2$ | $x^2 + x$ | $x^2 + x + 1$ |
| 3 | Cubic | $x^3$ | $x^3 + x$ | $x^3 + x + 1$ |
| 4 | Quartic | $x^4$ | $x^4 + x$ | $x^4 + x + 1$ |
| 5 | Quintic | $x^5$ | $x^5 + x$ | $x^5 + x + 1$ |
| 6 | Sextic or Hexic | $x^6$ | $x^6 + x$ | $x^6 + x + 1$ |
| 7 | Septic or Heptic | $x^7$ | $x^7 + x$ | $x^7 + x + 1$ |
| 8 | Octic | $x^8$ | $x^8 + x$ | $x^8 + x + 1$ |
| 9 | Nonic | $x^9$ | $x^9 + x$ | $x^9 + x + 1$ |
| 10 | Decic | $x^{10}$ | $x^{10} + x$ | $x^{10} + x + 1$ |

**Extra:**

What degree of polynomial is named "Hectic"? 100

Evaluate the polynomial $9x^2 + 15x + 4x^3 + x$ for $x = 2$. 100




LESSON 14-1

## Problem Solving

### Polynomials

**The table below shows expressions used to calculate the surface area and volume of various solid figures where *s* is side length, *l* is length, *w* is width, *h* is height, and *r* is radius.**

**Solid Figure Polynomials**

| Solid Figure | Surface Area | Volume |
|---|---|---|
| Cube | $6s^2$ | $s^3$ |
| Rectangular Prism | $2lw + 2lh + 2wh$ | $lwh$ |
| Right Cone | $\pi rl + \pi r^2$ | $\pi r^2 h$ |
| Sphere | $4\pi r^2$ | $\frac{4}{3}\pi r^3$ |

1. List the expressions that are trinomials.
   $2lw + 2lh + 2wh$
2. What is the degree of the expression for the surface area of a sphere?
   The degree of $4\pi r^2$ is 2.
3. A cube has side length of 5 inches. What is its surface area?
   150 square inches
4. If you know the radius and height of a cone, you can use the expression $(r^2 + h^2)^{0.5}$ to find its slant height. Is this expression a polynomial? Why or why not?
   No, its exponent is not a whole number.
5. If a sphere has a radius of 4 feet, what is its surface area and volume? Use $\frac{22}{7}$ for pi.
   $SA = 201.14$ ft$^2$; $V = 268.19$ ft$^3$

**Circle the letter of the correct answer.**

6. Which statement is true of all the polynomials in the volume column of the table?
   A They are trinomials
   B They are binomials.
   (C) They are monomials.
   D None of them are polynomials.
7. The height, in feet, of a baseball thrown straight up into the air from 6 feet above the ground at 100 feet per second after *t* seconds is given by the polynomial $-16t^2 + 100t + 6$. What is the height of the baseball 4 seconds after it was thrown?
   (F) 150 feet
   G 278 feet
   H 342 feet
   J 662 feet




CHAPTER 14-1

## Reading Strategies

### Focus on Vocabulary

A **monomial** is a number or a product of numbers and variables with exponents that are whole numbers. A **polynomial** is a monomial or the sum or difference of monomials. This chart can help you understand polynomials.

| Examples | Classifying Polynomials by the Number of Terms | Classifying Polynomials by their Degree |
|---|---|---|
| $3a^2b$ → monomial | monomial = 1 term | $15x + 9x^4 - 4x^2 + 20$ |
| $\frac{4x^2}{y^3}$ → not a monomial | binomial = 2 terms | highest power = degree = 4 |
| $6x2y - y3$ → polynomial | trinomial = 3 terms | $-13x^4 + x - 2x^6$ |
| $x^3 + 2y^{\frac{1}{2}}$ | | highest power = degree = 6 |

**Use the information in the chart to answer the following questions.**

1. What is a monomial?
   A number or a product of numbers and variables with exponents that are whole numbers.
2. What is a polynomial?
   One monomial or the sum or difference of monomials.
3. Explain why $x^3 + 2y^{\frac{1}{2}}$ is not a polynomial.
   The exponent of $y$ is not a whole number.
4. Rewrite $\frac{4x^2}{y^3}$ so that it does not have a variable in the denominator. Use your answer to explain why $\frac{4x^2}{y^3}$ is not a polynomial.
   $4x^2y^{-3}$; Negative numbers are not whole numbers, so the exponent of $y$ is not a whole number.
5. Classify the polynomial $-2x^2 + 5x + 7$ by the number of terms.
   trinomial
6. Classify the polynomial in Exercise 5 by its degree.
   degree 2




LESSON 14-1

## Puzzles, Twisters & Teasers

### Puzzle It Out!

**Complete the crossword puzzle using terms you learned in this lesson.**

**Across**

1. A polynomial with three terms is a trinomial.
5. An expression is not a polynomial if it has a variable in the denominator.
7. A letter that represents an unknown quantity is a variable.
8. The exponents in the terms of a polynomial must be whole numbers.
9. A polynomial with two terms is a binomial.

**Down**

2. The simplest type of polynomial is a monomial.
3. A number that precedes a variable is called the coefficient of the variable.
4. One monomial or the sum or difference of monomials is a polynomial.
5. You look at the exponents in a polynomial to determine its degree.
6. A trinomial is a polynomial with three terms.

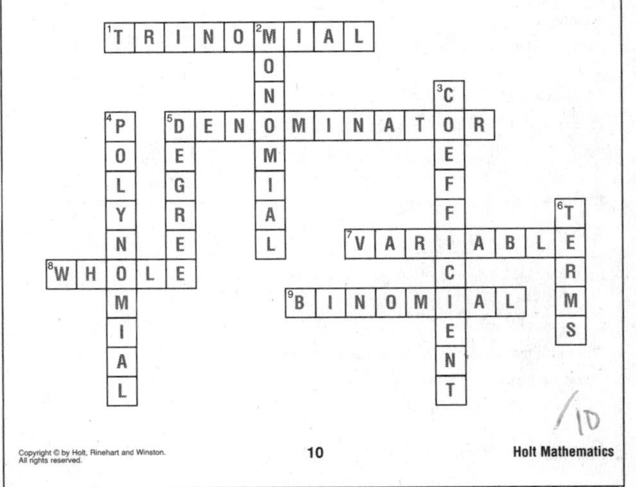

LESSON 14-2

## Practice A

### Simplifying Polynomials

**Identify the like terms in each polynomial.**

1. $x^2 - x + 3x^2 + 6x - 1$
   $x^2$ and $3x^2$, $-x$ and $6x$

2. $2t^2 + 7 - s^3 - t^2$
   $2t^2$ and $-t^2$

3. $2e^2 - 2 - 8e^2 + 3e - 2e^2$
   $2e^2$, $-8e^2$ and $-2e^2$

4. $2k - 3k^2 + 3k^2 - k$
   $-3k^2$ and $3k^2$, $2k$ and $-k$

5. $4ab^2 - 2b + 10 - ab^2$
   $4ab^2$ and $-ab^2$

6. $6z^2 + yz^2 + z^2 - 3yz^2$
   $6z^2$ and $z^2$, $yz^2$ and $-3yz^2$

7. $3g^2h^2 + 2g^2h - 7 + g^2h - 5g^2h^2$
   $3g^2h^2$ and $-5g^2h^2$, $2g^2h$, and $g^2h$

8. $m^3n^2 + 6 - m^2n^2 - 2m^3n^2 + 4m^2n^2$
   $m^3n^2$ and $-2m^3n^2$, $-m^2n^2$ and $4m^2n^2$

**Simplify.**

9. $k^2 - k + 2k^2 - 2k$
   $3k^2 - 3k$

10. $3x - 2 + 4x^2 + 6 - x^2$
    $3x^2 + 3x + 4$

11. $7r^2 - 7 - 23 - 3r^2$
    $4r^2 - 30$

12. $v^4 + v^2 + 2v^2 - 1$
    $v^4 + 3v^2 - 1$

13. $2(x + 2y)$
    $2x + 4y$

14. $3y + 2w - 7y + 5w$
    $-4y + 7w$

15. $4(2r - 2) - 10$
    $8r - 18$

16. $7 - w - 2 + 3w + 7$
    $2w + 12$

17. The height of a ball dropped off of a roof after $t$ seconds is given by the polynomial $-4(4t^2 - 10)$. Use the Distributive Property to write an equivalent expression.
    $-16t^2 + 40$




LESSON 14-2

## Practice B

### Simplifying Polynomials

**Identify the like terms in each polynomial.**

1. $x^2 - 8x + 3x^2 + 6x - 1$
   $x^2$ and $3x^2$; $-8x$ and $6x$

2. $2c^2 + d^3 + 3d^3 - 2c^2 + 6$
   $2c^2$ and $-2c^2$; $d^3$ and $3d^3$

3. $2x^2 - 2xy - 2y^2 + 3xy + 3x^2$
   $2x^2$ and $3x^2$; $-2xy$ and $3xy$

4. $2 - 9x + x^2 - 3 + x$
   $-9x$ and $x$; $2$ and $-3$

5. $xy - 5x + y - x + 10y - 3y^2$
   $-5x$ and $-x$; $y$ and $10y$

6. $6p + 2p^2 + pq + 2q^3 - 2p$
   $6p$ and $-2p$

7. $3a + 2b + a^2 - 5b + 7a$
   $3a$ and $7a$; $2b$ and $-5b$

8. $10m - 3m^2 + 9m^2 - 3m - m^3$
   $10m$ and $-3m$, $-3m^2$ and $9m^2$

**Simplify.**

9. $2h - 9hk + 6h - 6k$
   $8h - 9hk - 6k$

10. $9(x^2 + 2xy - y^2) - 2(x^2 + xy)$
    $7x^2 + 16xy - 9y^2$

11. $7qr - q^2r^3 + 2q^2r^3 - 6qr$
    $q^2r^3 + qr$

12. $8v^4 + 3v^2 + 2v^2 - 16$
    $8v^4 + 5v^2 - 16$

13. $3(x + 2y) + 2(2x - 3y)$
    $7x$

14. $7(1 - x) + 3x^2y + 7x - 7$
    $3x^2y$

15. $6(9y + 1) + 8(2 - 3y)$
    $30y + 22$

16. $a^2b - a^2 + ab^2 - 3a^2b + ab$
    $-2a^2b - a^2 + ab^2 + ab$

17. A student in Tracey's class created the following expression: $y^3 - 3y + 4(y^2 - y^3)$. Use the Distributive Property to write an equivalent expression.
    $-3y^3 + 4y^2 - 3y$




LESSON 14-2

## Practice C

### Simplifying Polynomials

**Identify the like terms in each polynomial.**

1. $x^2 - 18x + 3.5x^2 + 16x - 0.3$
   $x^2$ and $3.5x^2$; $-18x$ and $16x$

2. $s^3t + s^3 + 3s^3t - t^2 + 4st + 2s^3$
   $s^3t$ and $3s^3t$; $s^3$ and $2s^3$

3. $\frac{1}{3}g + \frac{1}{2}h + g^2 - \frac{5}{2}h$
   $\frac{1}{2}h$ and $-\frac{5}{2}h$

4. $\frac{1}{2}p^2 - 4p + 1 - p + \frac{9}{2}$
   $-4p$ and $-p$; $1$ and $\frac{9}{2}$

5. $12e^2 - 21ef - 8f^2 + 3ef - 22e^2$
   $12e^2$ and $-22e^2$;
   $-21ef$ and $3ef$

6. $2cd - c^2d^2 + 9c - 3c^2d^2 + c$
   $9c$ and $c$;
   $-c^2d^2$ and $-3c^2d^2$

7. $\frac{1}{5}mn - 3m^2n + \frac{3}{5}mn - mn^2$
   $\frac{1}{5}mn$ and $\frac{3}{5}mn$

8. $6m - m^5 + 2m^3 - 2m$
   $6m$ and $-2m$

**Simplify.**

9. $2k - 9kn - 2k - 2k + 5(n - k)$
   $-7k - 9kn + 5n$

10. $9(x^2 + xy - y^2) + \frac{1}{9}(x^2 - 9xy)$
    $9\frac{1}{9}x^2 + 8xy - 9y^2$

11. $\frac{7}{2}qr - q^2r^3 + \frac{1}{2}(q^2r^3 - 3qr)$
    $-\frac{1}{2}q^2r^3 + 2qr$

12. $\frac{8}{3}v^4 + v^2 + \frac{1}{2}(v^2 - 4)$
    $\frac{8}{3}v^4 + \frac{3}{2}v^2 - 2$

13. $4(x + 2y) - 2(2x - 3y) - 2$
    $14y - 2$

14. $7(1 - w) + 7(-2w + 3w^2 + 1) + 7$
    $21 - 21w + 21w^2$

15. $2(6 + 9q) - (6q - 9q^2 + 6q)$
    $9q^2 + 6q + 12$

16. $abc + a^2 + a^3b^2c - a^2 + abc$
    $2abc + a^3b^2c$

17. Mark's math teacher wrote the following expression on the board: $\frac{1}{2}\left(8 - 4y - \frac{1}{2}y^2\right) + \frac{1}{8}$. Use the Distributive Property to write an equivalent expression.
    $-2y - \frac{1}{4}y^2 + \frac{33}{8}$




LESSON 14-2

## Reteach

### Simplifying Polynomials

You can simplify a polynomial by combining like terms. Like terms have the same variables raised to the same powers. All constants are like terms.

$9 + 6y^3 - 8 + 7x^2y^3 + 3x^2y^3$

like terms    like terms

$7x^2y^3$ and $3x^2y^3$ both have the variable $x$ raised to the 2nd power and the variable $y$ raised to the 3rd power. Therefore, they are like terms.

**Identify the like terms in each polynomial**

1. $m + 3m^2 - 2m + 6 + 2m^2$
   $m$ and $-2m$; $3m^2$ and $2m^2$

2. $b - a^2b^2 - 2 + a^2 + 2a^2b^2$
   $2a^2b^2$ and $-a^2b^2$

3. $x^3 + 2 + 4x^3 - 9 + x$
   $x^3$ and $4x^3$; $2$ and $-9$

4. $9 + 4dg^2 + 4 + 6dg^2 + d^2$
   $4dg^2$ and $6dg^2$; $4$ and $9$

To simplify a polynomial, combine like terms. To combine like terms, add or subtract the coefficients. The variables and the exponents do not change.

$7x^2y^3 - 6y^3 + 3x^2y^3$

$7x^2y^3 + 6y^3 + 3x^2y^3$    Identify like terms.

$10x^2y^3 + 6y^3$    Combine coefficients of like terms.

$7 + 3 = 10$

**Simplify.**

5. $8a + 3ab^2 + 3a + 2ab^2$
   $5ab^2 + 11a$

6. $x^3 + 1 + 2x^3 + 3xy^2 - 3$
   $3x^3 + 3xy^2 - 2$

7. $y^4 + 2x^2y^3 - 3x^2 + 2y^4$
   $3y^4 + 2x^2y^3 - 3x^2$

LESSON 14-2

## Reteach

### Simplifying Polynomials (continued)

To simplify some polynomials, you first need to use the Distributive Property.

Distributive Property: $a(b + c) = a \cdot b + a \cdot c$

Simplify the polynomial.

$3(4y^2 + x^2) + 5x^2$

$3(4y^2 + x^2) + 5x^2 = (3 \cdot 4y^2) + (3 \cdot x^2) + 5x^2$ Apply the Distributive Property.

$= 12y^2 + 3x^2 + 5x^2$

$= 12y^2 + 3x^2 + 5x^2$ Combine like terms.

like terms

$= 12y^2 + 8x^2$

So, $3(4y^2 + x^2) + 5x^2$ simplified is $12y^2 + 8x^2$.

**Simplify.**

8. $2(2a^2 - b^2) + 3a^2$

$7a^2 - 2b^2$

9. $4(3x^2y + 2x) + 3x^2 + 5x$

$12x^2y + 3x^2 + 13x$

10. $-6mn + 3(m^4 + 3mn) - 2n^2$

$3m^4 + 3mn - 2n^2$

11. $8(y - 5) + 3y + 6y^2$

$6y^2 + 11y - 40$

12. $2(x^2 + 3x - 6) + 5(y + 2)$

$2x^2 + 6x + 5y - 2$

13. $4(xy^2 + 2y + 3) + 2(x^2 - 2xy^2)$

$8y + 12 + 2x^2$





LESSON 14-2

## Challenge

### Coming To Terms

**For each polynomial, use the simplified polynomial to find the missing term.**

| | Polynomial | Simplified Polynomial | Missing Term |
|---|---|---|---|
| 1. | $7x^2 - 4x + ? + 3x - 5$ | $5x^2 - x - 5$ | $-2x^2$ |
| 2. | $6(9x + ?)$ | $54x + 18$ | 3 |
| 3. | $12 + 2m + 3m^4 - 8m^2 + 5 + ? - 7m^2$ | $3m^4 - 15m^2 + 6m + 17$ | $4m$ |
| 4. | $?(2b^2 - 9b) + 4b^2 + 11b$ | $14b^2 - 34b$ | 5 |
| 5. | $7(x^3 + 3x) - 5x^3 + ?$ | $2x^3 + 11x$ | $-10x$ |
| 6. | $4ab + ? + 3a^2b^2 + 2ab - 8$ | $2a^2b^2 + 6ab - 8$ | $-a^2b^2$ |
| 7. | $? + 7 + 10w^2 - 4w^2 + 5w - 2$ | $6w^2 + 4w + 5$ | $-w$ |
| 8. | $2(t - 7) + ? - 2t^2$ | $-2t^2 + 14t - 14$ | $12t$ |
| 9. | $3hk - h^2k + hk^2 + 4hk + ? + 3hk^2$ | $4hk^2 - 3h^2k + 7hk$ | $-2h^2k$ |
| 10. | $4y^3 + 6y - 7y^2 + 2y^3 + ?$ | $6y^3 - 6y^2 + 6y$ | $y^2$ |
| 11. | $? - n^3 - \frac{1}{4}n^4 + \frac{1}{2}n^3 - \frac{1}{3}n^3$ | $\frac{1}{4}n^4 - \frac{5}{6}n^3$ | $\frac{1}{2}n^4$ |
| 12. | $3(12v^3 + ? + 2v^3) + v^3$ | $46v^3$ | $v^3$ |
| 13. | $1.5pq^3 + 0.7p^2q + ? + 2.4p^2q$ | $1.5pq^3 + 3.1p^2q - 3.8pq$ | $-3.8pq$ |
| 14. | $-9(? + 8w) + 4(-2w^2 + 18w)$ | $w^2$ | $-w^2$ |





LESSON 14-2

## Problem Solving

### Simplifying Polynomials

**Write the correct answer.**

1. The area of a trapezoid can be found using the expression $\frac{h}{2}(b_1 + b_2)$ where $h$ is height, $b$ is the length of base$_1$, and $b_2$ is the length of base$_2$. Use the Distributive Property to write an equivalent expression.

$\frac{hb_1}{2} + \frac{hb_2}{2}$

2. The sum of the measures of the interior angles of a polygon with $n$ sides is $180(n - 2)$ degrees. Use the Distributive Property to write an equivalent expression, and use the expression to find the sum of the measures of the interior angles of an octagon.

$180n - 360$; 1,080 degrees

3. The volume of a box of height $h$ is $2h^4 + h^3 + h^2 + h^2 + h$ cubic inches. Simplify the polynomial and then find the volume if the height of the box is 3 inches.

$2h^4 + h^3 + 2h^2 + h$;

210 cubic inches

4. The height, in feet, of a rocket launched upward from the ground with an initial velocity of 64 feet per second after $t$ seconds is given by $16(4t - t^2)$. Write an equivalent expression for the rocket's height after $t$ seconds. What is the height of the rocket after 4 seconds?

$64t - 16t^2$; 0 ft

**Circle the letter of the correct answer.**

5. The surface area of a square pyramid with base $b$ and slant height $l$ is given by the expression $b(b + 2l)$. What is the surface area of a square pyramid with base 3 inches and slant height 5 inches?

A 13 square inches

B 19 square inches

C 39 square inches

D 55 square inches

6. The volume of a box with a width of $3x$, a height of $4x - 2$, and a length of $3x + 5$ can be found using the expression $3x(12x^2 + 14x - 10)$. Which is this expression, simplified by using the Distributive Property?

F $36x^2 + 42x - 30$

G $15x^3 + 17x^2 - 7x$

H $36x^3 + 14x - 10$

J $36x^3 + 42x^2 - 30x$





LESSON 14-2

## Reading Strategies

### Understand Vocabulary

**Like terms** in a polynomial have the same variables raised to the same powers. After identifying like terms, you can combine them to simplify the polynomial.

| Identifying Like Terms | Simplifying Polynomials by Combining Like Terms |
|---|---|
| $7b + 4b^2 + 5 + 3b - b^2$ | Add or subtract the coefficients of like terms. |
| $7b$ and $3b$ are like terms. | Add the coefficients, 7 and 3, and keep the variable the same. $7b + 3b = 10b$ |
| $4b^2$ and $-b^2$ are like terms. | Subtract the coefficients, 4 and −1, and keep the variable the same. $4b^2 - b^2 = 3b^2$ |
| | $7b + 4b^2 + 5 + 3b - b^2 = 10b + 3b^2 + 5$ |

**Answer each question.**

1. Explain why $4b^2$ and $-b^2$ are like terms.

Both terms have the same variable raised to the same power ($b^2$).

2. Explain why $4x^3y^2$ and $15x^2y^2$ are not like terms.

The powers of $x$ in the two terms are different.

3. Describe a monomial that would be a like term for $10xy$. Then give an example.

The product of any number and $xy$: Possible answer: $4xy$

4. Identify the coefficient of each term in the polynomial $3c - 2c^2 + 10$.

The coefficient of $3c$ is 3 and the coefficient of $-2c^2$ is −2.

5. When you combine like terms, what do you do to the coefficients, and to the variables, and to the exponents of the terms?

Add the coefficients, but do not change the variables or the exponents.

6. In the polynomial $7x^3 - 4x^2 + x + 6$, what is the coefficient of $x$?

1

LESSON 14-2

## Puzzles, Twisters & Teasers

### Fix It!

**A magic square has the same sum for every column, row and diagonal. The sum is called the magic sum. In the magic square below, one of the polynomials is incorrect. Find the magic sum. Then, find the polynomial that does not fit, circle it, and FIX IT!**

| | | |
|---|---|---|
| $4x^2 - 2$ | $9x^2 - 7$ | $2x^2 - 6$ |
| $3x^2 - 9$ | $5x^2 - 5$ | $7x^2 - 1$ |
| $8x^2 - 4$ | $10x^2 - 2$ | $6x^2 - 8$ |

Magic sum $15x^2 - 15$

Incorrect polynomial $10x^2 - 2$

Corrected polynomial $x^2 - 3$

LESSON 14-3

## Practice A

### Adding Polynomials

**Add.**

1. $(3a + 3) + (5a + 2)$ — $8a + 5$
2. $(4x - x^2 - 9) + (2 - 2x + 6x^2)$ — $5x^2 + 2x - 7$
3. $(5y + 2y^2 + 3) + (-5y^2 + 2y)$ — $-3y^2 + 7y + 3$
4. $(3k^3 + 7k^2 + k) + (k^3 - 2k^2 - 10)$ — $4k^3 + 5k^2 + k + 10$
5. $(2wv - 2w^2v) + (7wv^2 + 6w^2v) + (-3wv + wv^2)$
   $4w^2v + 8wv^2 - wv$
6. $(b^2c^2 - b^2c + 3bc) + (b^2c^2 - bc + 1) + (2b^2c - bc - 1)$
   $2b^2c^2 + b^2c + bc$
7. $(e^2 + 3e + 2) + (9 - 6e + e^2) + (9e - 2 - e^2) + (4e^2 - 7e - 8)$
   $5e^2 - e + 1$
8. $(f^4 - 2f + f^3 - 4) + (3f^3 + 3) + (f^4 - f) + (3 - f^3 + f^4 - f) + (4 - f)$
   $3f^4 + 3f^3 - 5f + 6$
9. Micah is putting a frame of width $w$ around an 18-inch by 24-inch poster. Find an expression for the length of framing material she needs.
   $4w + 84$
10. Each side of an equilateral triangle has length $w + 3$. Each side of a square has length $2w + 2$. Write an expression for the sum of the perimeter of the equilateral triangle and the perimeter of the square.
   $11w + 17$




LESSON 14-3

## Practice B

### Adding Polynomials

**Add.**

1. $(a^2 + a + 3) + (15a^2 + 2a + 9)$ — $16a^2 + 3a + 12$
2. $(5x + 2x^2) + (3x - 2x^2)$ — $8x$
3. $(mn - 10 + mn^2) + (5 + 3mn - 4mn^2)$ — $-3mn^2 + 4mn - 5$
4. $(7y^2z + 9 + yz^2) + (y^2z - 2yz^2)$ — $8y^2z - yz^2 + 9$
5. $(s^3 + 3s - 3) + (2s^3 + 9s - 2) + (s - s^3)$
   $2s^3 + 13s - 5$
6. $(6wv - 4w^2v + 7wv^2) + (5w^2v - 7wv^2) + (wv^2 - 5wv + 6w^2v)$
   $7w^2v + wv^2 + wv$
7. $(6b^2c^2 - 4b^2c + 3bc) + (9b^2c^2 - 4bc + 12) + (2b^2c - 3bc - 8)$
   $15b^2c^2 - 2b^2c - 4bc + 4$
8. $(7e^2 + 3e + 2) + (9 - 6e + 4e^2) + (9e + 2 - 6e^2) + (4e^2 - 7e + 8)$
   $9e^2 - e + 21$
9. $(f^4g - fg^3 + 2fg - 4) + (3fg^3 + 3) + (4f^4g - 5fg) + (3 - 12fg^3 + f^4g)$
   $6f^4g - 10fg^3 - 3fg + 2$
10. Six blocks of height $4h + 4$ each and 3 blocks of height $8 - 2h$ each are stacked on top of each other to form one big tower. Find an expression for the overall height of the tower.
   $18h + 48$

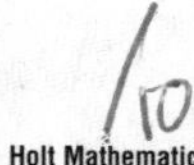




LESSON 14-3

## Practice C

### Adding Polynomials

**Add.**

1. $(3a^2 + 3a + 3) + (15a^2 + 2a - 9)$ — $18a^2 + 5a - 6$
2. $(3g^2 + 2hg) + (5g^2 + hg - 9h^2g)$ — $8g^2 - 9h^2g + 3hg$
3. $(5x - 4y + 6y^2) + (-5y^2 + 2y - 3x)$ — $y^2 + 2x - 2y$
4. $(m^2 + 3m - 2) + (4m^2 + m - 9)$ — $5m^2 + 4m - 11$
5. $\left(\frac{3}{2}s^3 + \frac{1}{6}s^2 + \frac{1}{3}t\right) + \left(\frac{1}{2}s^3 + \frac{1}{3}t\right)$
   $2s^3 + \frac{1}{6}s^2 + \frac{2}{3}t$
6. $(6wv - 4w^2v + 7wv^2) + (5w^2v - 7wv^2) + (wv^2 - 5wv + 6w^2v) - 7w^2v$
   $wv^2 + wv$
7. $(6b^2c^2 - 11b^2c + 3.25bc) + (9b^2c^2 - 45bc + 12) + (2b^2c - 3.25bc - 8)$
   $15b^2c^2 - 9b^2c - 45bc + 4$
8. $\left(\frac{1}{7}e^2 + 3e + 2\right) + \left(9 - 6e + \frac{1}{4}e^2\right) + \left(9e - 2 + \frac{6}{7}e^2\right) + (4e^2 - 7e - 8)$
   $5\frac{1}{4}e^2 - e + 1$
9. $\left(\frac{1}{2}f^4g - 2fg^3 + \frac{1}{2}fg - 4\right) + (3fg^3 + 3) + \left(\frac{2}{4}f^4g - \frac{5}{2}fg\right) + (3 - 12fg^3 + f^4g)$
   $2f^4g - 11fg^3 - 2fg + 2$
10. The area of Kenny's back yard is $2h^2 - 9h + 8$ square yards. The area of Kenny's front yard is $h^2 - 38$ square yards. Write an expression for the total area of Kenny's back and front yards.
   $3h^2 - 9h - 30$ square yards

LESSON **14-3**

## Reteach

### Adding Polynomials

Adding polynomials is like simplifying polynomials.
You can regroup the terms and then combine like terms. Or you can place the polynomials in columns and then combine like terms.

Find an expression for the perimeter of the triangle below.

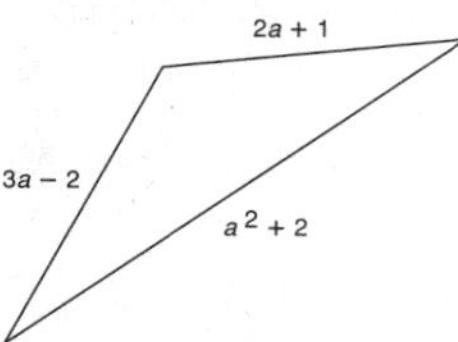


To find an expression for the perimeter, add the polynomials.

$(2a + 1) + (3a - 2) + (a^2 + 2)$

like terms    like terms

Place like terms in columns and combine them.

$$\begin{array}{r} 2a + 1 \\ 3a - 2 \\ + a^2 \quad\quad + 2 \\ \hline a^2 + 5a + 1 \end{array}$$

So, an expression for the perimeter of the triangle is $a^2 + 5a + 1$.

**Add.**

1. $(3x^2 + 3xy^3 + 5y + 2) + (4xy^3 - 3y)$

$$\begin{array}{r} 3x^2 + 3xy^3 + 5y + 2 \\ + \quad\quad 4xy^3 - 3y \quad\quad \\ \hline \end{array}$$

$3x^2 + 7xy^3 + 2y + 2$

2. $(4a^2b - 3a^2 + 3b) + (6a^2b + 4ab - 2b)$

$$\begin{array}{r} 4a^2b - 3a^2 \quad\quad + 3b \\ + 6a^2b \quad\quad + 4ab - 2b \\ \hline \end{array}$$

$10a^2b - 3a^2 + 4ab + b$

3. $(4mn + 5n^3 + 3n) + (3m^2 + 5n)$

$4mn + 3m^2 + 5n^3 + 8n$

4. $(-5r^3 + 2r + 7) + (2r^3 + 4r^2 - 6r + 1)$

$-3r^3 + 4r^2 - 4r + 8$




LESSON **14-3**

## Challenge

### Perimeters

**Write a polynomial for the perimeter of each figure. Simplify each polynomial.**

1.

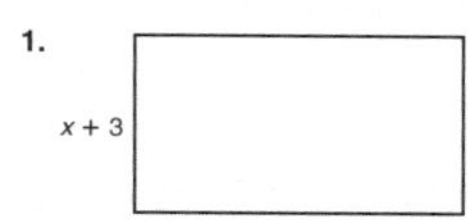


Perimeter: $4x + 20$

2.

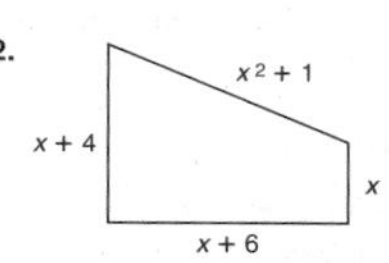


Perimeter: $x^2 + 3x + 11$

3.

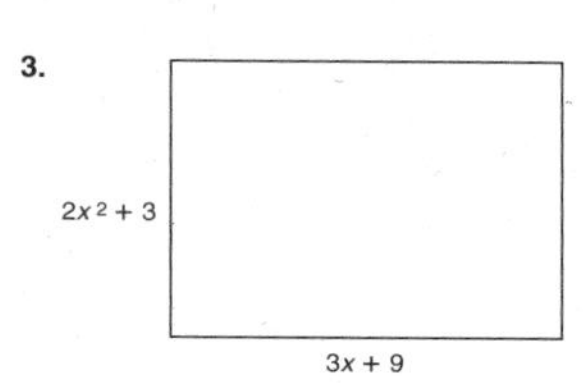


Perimeter: $4x^2 + 6x + 24$

4. $x^2 + 3$, $x^2 + x + 2$, $x^2$

Perimeter: $3x^2 + x + 5$

5.

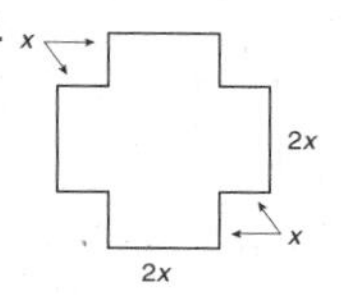


Perimeter: $16x$

6. $2x + 1.4$, $3x$, $3x$, $2x + 1.4$

Perimeter: $10x + 2.8$

LESSON **14-3**

## Problem Solving

### Adding Polynomials

**Write the correct answer.**

1. What is the perimeter of the quadrilateral?

$3x + 1$, $4x - 4$, $x^2 - 3$, $x^2 + 2x - 1$

$2x^2 + 9x - 7$

2. Jasmine purchased two rugs. One rug covers an area of $x^2 + 8x + 15$ and the other rug covers an area of $x^2 + 3x$. Write and simplify an expression for the combined area of the two rugs.

$2x^2 + 11x + 15$

3. Anita's school photo is 12 inches long and 8 inches wide. She will surround the photo with a mat of width $w$. She will surround the mat with a frame that is twice the width of the mat. Find an expression for the perimeter of the framed photo.

$24w + 40$

4. The volume of a right cylinder is given by $\pi r^2h$. The volume of a right cone is given by $\frac{1}{3}\pi r^2h$. Write and simplify an expression for the total volume of a right cylinder and right cone combined, if the cylinder and cone have the same radius and height. Use 3.14 for $\pi$.

about $4.19r^2h$

**Choose the letter of the correct answer.**

5. Each side of a square has length $4s - 2$. Which is an expression for the perimeter of the square.

A $8s - 2$
(B) $16s - 8$
C $8s - 4$
D $16s - 4$

6. The side lengths of a certain triangle can be expressed using the following binomials: $x + 3$, $2x + 2$, and $3x - 2$. Which is an expression for the perimeter of the triangle?

F $2x + 5$
G $2x - 1$
H $3x + 5$
(J) $6x + 3$

7. What polynomial can be added to $2x^2 + 3x + 1$ to get $2x^2 + 8x$?

A $5x$
B $5x + 1$
C $5x^2 - 1$
(D) $5x - 1$

8. Which of the following sums is NOT a binomial when simplified?

(F) $(b^2 + 5b + 1) + (b^2 + 5b + 1)$
G $(b^2 + 5b + 1) + (b^2 + 5b - 1)$
H $(b^2 + 5b + 1) + (b^2 - 5b + 1)$
J $(b^2 + 5b + 1) + (-b^2 + 5b + 1)$




LESSON **14-3**

## Reading Strategies

### Follow a Procedure

There are two ways to add polynomials.

| Adding Polynomials Horizontally | Adding Polynomials Vertically |
|---|---|
| **Step 1:** The Associative Property of Addition allows you to regroup the terms.<br>**Step 2:** Identify like terms.<br>**Step 3:** Combine like terms by adding their coefficients.<br>$(7x^3 + 5x^2 - 8) + (6x^2 + 20)$<br>$7x^3 + 5x^2 - 8 + 6x^2 + 20$<br>$7x^3 + (5x^2) - [8] + (6x^2) + [20]$<br>$7x^3 + 11x^2 + 12$ | **Step 1:** Identify like terms.<br>**Step 2:** Place like terms in columns. If you rearrange terms, remember to keep the correct sign with each term.<br>**Step 3:** Combine like terms by adding their coefficients.<br>$(2h^2 + 9h^3 + 7) + (11h - 4h^3 - 3h^2)$<br>$9h^3 + 2h^2 \quad + 7$<br>$-4h^3 - 3h^2 + 11h$<br>$5h^3 - h^2 + 11h + 7$ |

**Use the information in the chart to answer the following questions.**

1. What property allows you to regroup terms in an addition problem?

The Associative Property of Addition.

2. How do you combine like terms?

by adding their coefficients

3. What happens to a term that has no like terms with which to combine it?

It is unchanged when written in the simplified polynomial

4. What are the first steps when adding polynomials vertically?

Identify like terms and place them in columns.

5. How would you rearrange $7 - 4x^2 - 6x + 5x^3$ so that you can vertically add it to $10x^3 + 3x^2 - 9x + 1$?

$5x^3 - 4x^2 - 6x + 7$

6. When placing like terms in columns to add polynomials vertically, when do you skip spaces?

When a term in one polynomial does not have a like term in the other polynomial.

## LESSON 14-3 Puzzles, Twisters & Teasers

### *It's Amazing!*

**Make your way through the maze by finding the sum of each set of polynomials. You may move up, down, left, right or diagonally, but you may not enter a square more than once.**

Sum

1. $(5x^2 - 4x) + (3x^2 + 7x - 2)$ $\underline{8x^2 + 3x - 2}$
2. $(-6x^2 + x - 8) + (11x^2 - 5x)$ $\underline{5x^2 - 4x - 8}$
3. $(2x^2 + 12x - 14) + (-8x^2 - 15x) + (7x + 20)$ $\underline{-6x^2 + 4x + 6}$
4. $\begin{array}{r} 10x^2 - 4x + 17 \\ +\ 3x^2 + 9x - 20 \end{array}$ $\underline{13x^2 + 5x - 3}$
5. $\begin{array}{r} 6x^3 - 3x^2 + 7x + 4 \\ +\ 5x^3 + 11x^2 \qquad - 10 \end{array}$ $\underline{11x^3 + 8x^2 + 7x - 6}$

| $8x^2$ | – | $3x$ | $5x^2$ | + | $4x^2$ | + | 9 | – | + |
|---|---|---|---|---|---|---|---|---|---|
| + | $2x$ | 2 | $17x^2$ | – | $6x$ | + | $3x^2$ | 5 | + |
| $7x$ | $3x$ | – | + | $12x^2$ | $4x$ | 13 | + | $8x$ | $6x$ |
| + | + | $4x^2$ | 18 | 12 | – | 6 | $11x^3$ | + | $11x^3$ |
| $13x$ | 2 | – | $-6x^2$ | 8 | 9 | 3 | – | 7 | $4x^2$ |
| $-5x^2$ | $4x$ | + | $6x$ | $13x^2$ | – | $5x$ | – | $6x^2$ | $-8x^2$ |
| $8x^2$ | $4x^2$ | + | 6 | + | 13 | $9x$ | $6x^2$ | $8x^2$ | + |
| – | – | 3 | + | $5x$ | $13x$ | $14x$ | + | $7x^2$ | $7x$ |
| $-2x^2$ | $11x^2$ | + | + | – | 37 | $11x^3$ | – | $5x^2$ | – |
| 8 | $5x$ | $8x$ | $8x^2$ | + | 3 | $11x^2$ | $14x^2$ | + | 6 |

## LESSON 14-4 Practice A

### *Subtracting Polynomials*

**Find the opposite of each polynomial.**

1. $xy^2$ $\underline{-xy^2}$
2. $-8p^4q^3$ $\underline{8p^4q^3}$
3. $5 - 12a$ $\underline{12a - 5}$
4. $4k^3j + 3k$ $\underline{-4k^3j - 3k}$
5. $2u^2 - 6u + 9$ $\underline{-2u^2 + 6u - 9}$
6. $-d^4e^3 - 2d^3e^4 - 8de$ $\underline{d^4e^3 + 2d^3e^4 + 8de}$

**Subtract.**

7. $(b^2 + 9) - (3b^2 - 5)$ $\underline{-2b^2 + 14}$
8. $(2x^2 - 3x) - (x^2 + 7)$ $\underline{x^2 - 3x - 7}$
9. $(-2x^2 + 8x + 9) - (9x^2 + 6x - 2)$ $\underline{-11x^2 + 2x + 11}$
10. $(5y^3 + 8y^2 + y) - (9y^3 + 7y^2 - 2)$ $\underline{-4y^3 + y^2 + y + 2}$
11. $(5q^5 + q^4 + q^3 + 2q^2 - 5q + 3) - (3 - 2q^3 + q^2 - 10q)$
    $\underline{5q^5 + q^4 + 3q^3 + q^2 + 5q}$
12. $-2f - (f^4 + 3f^3g + 2f^2g - 3fg - 2f - 10)$
    $\underline{-f^4 - 3f^3g - 2f^2g + 3fg + 10}$
13. $(nm^2 - 2n^3 + 4n^2m^2 - nm + 15) - (-10nm + 4n^2 - 5nm^2 + 6n^2m^2 - 2n^3)$
    $\underline{-2n^2m^2 + 6nm^2 - 4n^2 + 9nm + 15}$
14. Suppose the number of boxes (in millions) of crayons manufactured annually by the Great Crayon Company is shown by the expression $9x^2 - 8x + 7$. If the rest of the crayon-making companies manufacture $6x^2 - 3x + 4$ crayons, how many more crayons does The Great Crayon Company manufacture each year than the rest of the crayon-making companies?
    $\underline{3x^2 - 5x + 3 \text{ crayons}}$




## LESSON 14-4 Practice B

### *Subtracting Polynomials*

**Find the opposite of each polynomial.**

1. $18xy^3$ $\underline{-18xy^3}$
2. $-9a + 4$ $\underline{9a - 4}$
3. $6d^2 - 2d - 8$ $\underline{-6d^2 + 2d + 8}$

**Subtract.**

4. $(4n^3 - 4n + 4n^2) - (6n + 3n^2 - 8)$ $\underline{4n^3 + n^2 - 10n + 8}$
5. $(-2h^4 + 3h - 4) - (2h - 3h^4 + 2)$ $\underline{h^4 + h - 6}$
6. $(6m + 2m^2 - 7) - (-6m^2 - m - 7)$ $\underline{8m^2 + 7m}$
7. $(17x^2 - x + 3) - (14x^2 + 3x + 5)$ $\underline{3x^2 - 4x - 2}$
8. $w + 7 - (3w^4 + 5w^3 - 7w^2 + 2w - 10)$
   $\underline{-3w^4 - 5w^3 + 7w^2 - w + 17}$
9. $(9r^3s - 3rs + 4rs^3 + 5r^2s^2) - (2rs^2 - 2r^2s^2 + 6rs + 7r^3s - 9)$
   $\underline{2r^3s + 7r^2s^2 + 4rs^3 - 2rs^2 - 9rs + 9}$
10. $(3qr^2 - 2 + 14q^2r^2 - 9qr) - (-10qr + 11 - 5qr^2 + 6q^2r^2)$
    $\underline{8q^2r^2 + 8qr^2 + qr - 13}$
11. The volume of a rectangular prism, in cubic meters, is given by the expression $x^3 + 7x^2 + 14x + 8$. The volume of a smaller rectangular prism is given by the expression $x^3 + 5x^2 + 6x$. How much greater is the volume of the larger rectangular prism?
    $\underline{2x^2 + 8x + 8 \text{ cubic meters}}$
12. Sarah has a table with an area, in square inches, given by the expression of $y^2 + 30y + 200$. She has a tablecloth with an area, in square inches, given by the expression of $y^2 + 18y + 80$. She wants the tablecloth to cover the top of the table. What expression represents the number of square inches of additional fabric she needs to cover the top of the table?
    $\underline{12y + 120 \text{ more square inches of fabric}}$




## LESSON 14-4 Practice C

### *Subtracting Polynomials*

**Find the opposite of each polynomial.**

1. $\frac{1}{8}c^5d^4e$ $\underline{-\frac{1}{8}c^5d^4e}$
2. $-1.9f + 4g - 2.8h^4$ $\underline{1.9f - 4g + 2.8h^4}$
3. $mn^2 + mn - m^2n$ $\underline{-mn^2 - mn + m^2n}$

**Subtract.**

4. $\left(k^3 - \frac{1}{4}k + \frac{3}{4}k^2\right) - \left(\frac{1}{2}k + \frac{3}{4}k^2 - 6\right)$
   $\underline{k^3 + \frac{3}{4}k + 6}$
5. $(100m^2 - 25m^3 + 35) - (94m^2 + 30m + 5m^3 - 25)$
   $\underline{-30m^3 + 6m^2 - 30m + 60}$
6. $(17n^3p - 12np + 4np^3 + 19) - (12np^3 + 6np + 7n^3p - 9)$
   $\underline{10n^3p - 8np^3 - 18np + 28}$
7. $13q^5r^5 - \left(\frac{1}{25}q^3 + 2r^6 - 10\right) - 11q^5r^5$
   $\underline{2q^5r^5 - \frac{1}{25}q^3 - 2r^6 + 10}$
8. $(3st^2 - 2s^3 + 14s^2t^2) - (+4s^2 - 5st^2 + 6s^2t^2) - (-3s^3 + 14s^2)$
   $\underline{8s^2t^2 + s^3 + 8st^2 - 18s^2}$
9. The area of a square, in square yards, is given by the expression $4u^4 + 8u^3 + 12u^2 + 8u + 4$. The area of a smaller square is given by the expression $4u^4 + 4u^3 + 5u^2 + 2u + 1$. How much greater is the area of the larger square?
   $\underline{4u^3 + 7u^2 + 6u + 3 \text{ square yards}}$
10. The volume of a rectangular prism, in cubic meters, is given by the expression $c^3 + 4c^2 + 3c$. The volume of a smaller rectangular prism is given by the expression $c^3 - c^2 - 2c$. How much greater is the volume of the larger rectangular prism?
    $\underline{5c^2 + 5c \text{ cubic meters}}$

LESSON 14-4

## Reteach
### Subtracting Polynomials

When subtracting polynomials, you can distribute a factor of −1.

Subtract. $(5x^2 + 7x + 3) - (4x^2 + 3x - 5)$.

Rewrite the expression. $(5x^2 + 7x + 3) + (-1)(4x^2 + 3x - 5)$.

Apply the Distributive Property.

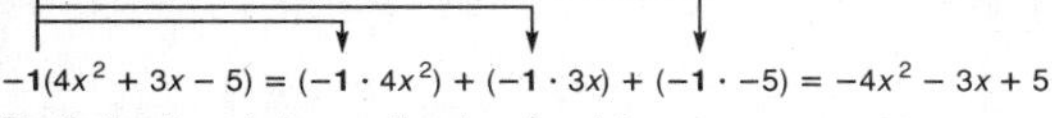

$-\mathbf{1}(4x^2 + 3x - 5) = (-\mathbf{1} \cdot 4x^2) + (-\mathbf{1} \cdot 3x) + (-\mathbf{1} \cdot -5) = -4x^2 - 3x + 5$

Distributing the −1 changes the sign of each term.
$(5x^2 + 7x + 3) + (-4x^2 - 3x + 5)$

Use the Associative Property to remove parentheses and combine like terms.

$5x^2 + 7x + 3 - 4x^2 - 3x + 5 = x^2 + 4x + 8$

**Subtract.**

1. $(3b^3 + 4b^2 + 6) - (b^3 - 5b - 3)$

| | |
|---|---|
| $3b^3 + 4b^2 + 6 + -1(b^3 - 5b - 3)$ | Rewrite the expression. |
| $3b^3 + 4b^2 + 6 + (-b^3 + 5b + 3)$ | Apply the Distributive Property. |
| $3b^3 + 4b^2 + 6 - b^3 + 5b + 3$ | Remove the parentheses. |

$2b^3 + 4b^2 + 5b + 9$

2. $(3m^2n^2 - 4m^2n + m^2) - (m^2n^2 + 5m^2n - 5)$

$2m^2n^2 - 9m^2n + m^2 + 5$

3. $(2x^3y^2 + x^2y - 4) - (x^2y - 8x + 3)$

$2x^3y^2 + 8x - 7$

4. $(6y^2 + 3xy - 9x^2) - (-4y^2 + 8xy + x^2)$

$10y^2 - 5xy - 10x^2$

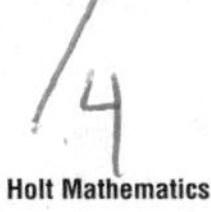




LESSON 14-4

## Challenge
### Grade a Polynomial Quiz

**Check the student's quiz shown below. If an answer is wrong, write the correct answer.**

Name: Poly Nomial Date: October 5, 2003

**DIRECTIONS: Find each sum. Show all your work.**

1. $(2 - 3x + x^2) + (-5 + 7x - 3x^2 + x^3)$

$x^2 - 3x + 2$
$+ x^3 - 3x^2 + 7x - 5$
$x^3 - 2x^2 + 10x - 3$

$x^3 - 2x^2 + 4x - 3$

2. $(-5b^3 + 6b^2 - 1) + (4b^3 + 3b^2 + 2)$

$-5b^3 + 6b^2 - 1$
$+ 4b^3 + 3b^2 + 2$
$-b^3 + 9b^2 + 1$

Correct

3. $(3m^3 - 4m^2 - 7 + m) + (7m^2 - 4m + 3)$

$3m^3 - 4m^2 + m - 7$
$+ 7m^2 - 4m + 3$
$3m^3 + 3m^2 - 3m - 4$

Correct

4. $(-8t + 6t^3 - 1 + 4t^2) + (t^3 - 6t^2 + t - 1)$

$6t^3 + 4t^2 - 8t - 1$
$+ t^3 - 6t^2 + t - 1$
$7t^3 - 2t^2 + 7t - 2$

$7t^3 - 2t^2 - 7t - 2$

DIRECTIONS: Find each difference. Show all your work.

5. $(7 + p - 5p^3 + 2p^2) - (-5p^3 + 3p - 7)$

$-5p^3 + 2p^2 + p + 7$
$-5p^3 + 3p - 7$
$-10p^3 + 2p^2 + 4p$

$2p^2 - 2p + 14$

6. $(-a^3 + 3a^2 - 4 + 2a) - (3a^3 + a^2 + 2)$

$-a^3 + 3a^2 + 2a - 4$
$- 3a^3 - a^2 - 2$
$-4a^3 + 2a^2 + 2a - 6$

Correct

7. $(-4h^4 + h^2 - 9h + h^3) - (4h^2 - 3h^4 - 9h)$

$4h^4 + h^3 + h^2 - 9h$
$- 3h^4 + 4h^2 - 9h$
$h^4 + h^3 + 5h^2 - 18h$

$h^4 + h^3 - 3h^2$

8. $(-8k + 3k^3 - 6 + k^2) - (5k^3 - k^2 + k^4 - 8)$

$3k^3 + k^2 - 8k - 6$
$- 5k^3 + k^2 - k - 8$
$-3k^3 + 2k^2 - 9k - 14$

$-k^4 - 2k^3 + 2k^2 - 8k + 2$




LESSON 14-4

## Problem Solving
### Subtracting Polynomials

**Write the correct answer.**

1. Molly made a frame for a painting. She cut a rectangle with an area of $x^2 + 3x$ square inches from a piece of wood that had an area of $2x^2 + 9x + 10$ square inches. Write an expression for the area of the remaining frame.

$x^2 + 6x + 10$

2. The volume of a rectangular prism, in cubic inches, is given by the expression $2t^3 + 7t^2 + 3t$. The volume of a smaller rectangular prism is given by the expression $t^3 + 2t^2 + t$. How much greater is the volume of the larger rectangular prism?

$t^3 + 5t^2 + 2t$

3. The area of a square piece of cardboard is $4y^2 - 16y + 16$ square feet. A piece of the cardboard with an area of $2y^2 + 2y - 12$ square feet is cut out. Write an expression to show the area of the cardboard that is left.

$2y^2 - 18y + 28$

4. A container is filled with $3a^3 + 10a^2 - 8a$ gallons of water. Then $2a^3 - 3a^2 - 3a + 2$ gallons of water are poured out. How much water is left in the container?

$a^3 + 13a^2 - 5a - 2$ gallons

**Circle the letter of the correct answer.**

5. The perimeter of a rectangle is $4x^2 + 2x - 2$ meters. Its length is $x^2 + x - 2$ meters. What is the width of the rectangle?

A $3x^2 + x + 2$ meters
B $2x^2 + 2$ meters
(C) $x^2 + 1$ meters
D $\frac{3}{2}x - \frac{1}{2}x + 1$ meters

6. On a map, points A, B, and C lie in a straight line. Point A is $x^2 + 2xy + 5y$ miles from Point B. Point C is $3x^2 - 5xy + 2y$ miles from Point A. How far is Point B from Point C?

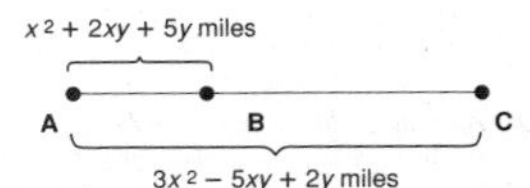


F $-2^2 + 7 + 3y$ miles
G $4x^2 - 3xy + 7y$ miles
H $-4x^2 + 3xy - 7y$ miles
(J) $2x^2 - 7xy - 3y$ miles




CHAPTER 14-4

## Reading Strategies
### Compare and Contrast

You can subtract polynomials horizontally or vertically.

| Subtracting Polynomials Horizontally | Subtracting Polynomials Vertically |
|---|---|
| **Step 1:** Rewrite the problem as a sum of the first polynomial and the opposite of the second polynomial. | **Step 1:** Rewrite the problem as a sum of the first polynomial and the opposite of the second polynomial. |
| **Step 2:** The Associative Property allows you to regroup the terms. | **Step 2:** Identify like terms. |
| **Step 3:** Identify like terms. | **Step 3:** Place like terms in columns. If you rearrange terms, remember to keep the correct sign with each term. |
| **Step 4:** Combine like terms by adding their coefficients. | **Step 4:** Combine like terms by adding their coefficients. |
| $(\mathbf{2x^3 + 9x^2 + 18}) - (\mathbf{5x^2 + 20})$ | $(\mathbf{2h^2 + 9h^3 + 7}) - (\mathbf{11h - 4h^3 - 3h^2})$ |
| $(2x^3 + 9x^2 + 18) + (-5x^2 - 20)$ | $(2h^2 + 9h^3 + 7) + (-11h + 4h^3 + 3h^2)$ |
| $2x^3 + 9x^2 + 18 - 5x^2 - 20$ | $9h^3 + 2h^2 + 7$ |
| $2x^3 + (9x^2) + [18] - (5x^2) - [20]$ | $4h^3 + 3h^2 - 11h$ |
| $2x^3 + 4x^2 - 2$ | $13h^3 + 5h^2 - 11h + 7$ |

**Use the information in the chart to answer the following questions.**

1. What is the first step when subtracting polynomials? Rewrite the problem as a sum of the first polynomial and the opposite of the second polynomial.

2. How do you find the opposite of a polynomial?
by changing the sign of each of the terms in the polynomial

3. Compare the two procedures for subtracting polynomials. How are they the same? How are they different? Same: In both processes you rewrite the problem as a sum of the first polynomial and the opposite of the second polynomial and you add like terms. Different: In the vertical process, you rewrite the problem aligning the like terms.

4. Compare the process of subtracting polynomials to that of adding polynomials. How are they the same? How are they different?
Same: You add like terms in both processes. Different: To subtract, you rewrite the problem as a sum of the first polynomial and the opposite of the second polynomial and you add like terms.

LESSON 14-4

## Puzzles, Twisters & Teasers

### Digital Displays!

How many home runs did Hank Aaron hit in his baseball career? 755

To discover the answer, find the missing terms in each of the solutions to the following problems. Then, shade in the sections of the digital display that contain the missing terms. Read the answer from the display.

1. $(7x^2 + 4x - 5) - (9x + 3) = 7x^2 - 5x -$ 8
2. $(4x^2 - 10x + 17) - (8x + 7) = 4x^2 -$ $18x$ $+$ 10
3. $(-4x^2 + 6) - (-5x^2 + 3x - 1) =$ $x^2$ $- 3x +$ 7
4. $(3x^3 + 14x^2 + x - 7) - (4x^2 + 6x - 13) =$ $3x^3$ $+ 10x^2 - 5x +$ 6
5. $(6x^2 + 11) - (x^2 + 4) - (2x^2 - 6) = 3x^2 +$ 13
6. $(-2x^3 + 7x - 4) - (-6x^3 - 3x^2 + 2x - 5) =$ $4x^3$ $+$ $3x^2$ $+ 5x +$ 1
7. $(25x^3 + 3) - (6x^2 + 11x) = 25x^3 -$ $6x^2$ $-$ $11x$ $+ 3$

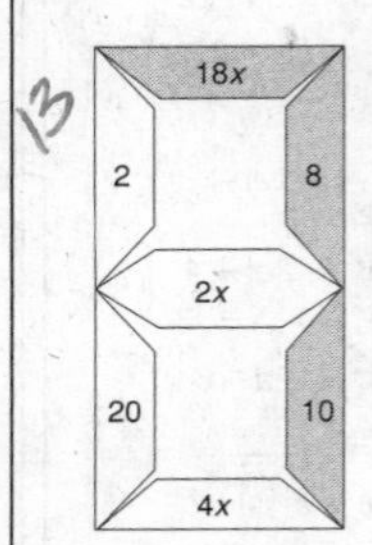


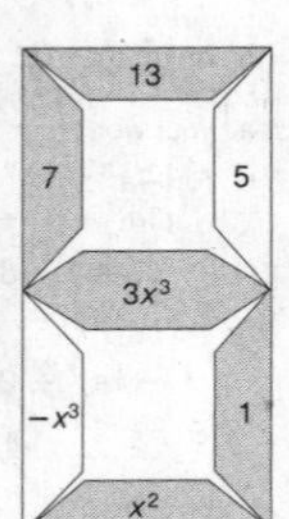


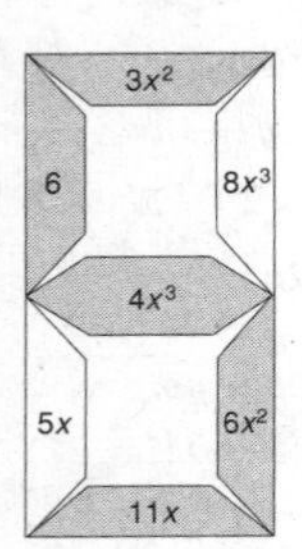

/22




LESSON 14-5

## Practice A

### Multiplying Polynomials by Monomials

Multiply.

1. $(x^2)(-y^3)$ — $-x^2y^3$
2. $(-9z^4)(-z^2)$ — $9z^6$
3. $(-3a^3b^2)(-ab^2)$ — $3a^4b^4$
4. $(-3hi^2)(3h^2i^2)$ — $-9h^3i^4$
5. $2(q^2 - 8)$ — $2q^2 - 16$
6. $-x(4x^4 - 12)$ — $-4x^5 + 12x$
7. $5y(3y^2 + 2y)$ — $15y^3 + 10y^2$
8. $6z\left(-\frac{1}{3}a^5 + 2a\right)$ — $-2a^5z + 12az$
9. $-2jk(4jk + 2j - 2k + 10)$ — $-8j^2k^2 - 4j^2k + 4jk^2 - 20jk$
10. $6mn(-m^2 - n^2 + 2mn)$ — $-6m^3n - 6mn^3 + 12m^2n^2$
11. $-pq^2(p^2q^2 + 2p^2q + 11)$ — $-p^3q^4 - 2p^3q^3 - 11pq^2$
12. $4r^4(r^2 - 2r + 1)$ — $4r^6 - 8r^5 + 4r^4$
13. $4s(-s^2 - 2t^2 + 3)$ — $-4s^3 - 8st^2 + 12s$
14. $7u^2(3u^2v + 7u^2 - 2u + 1)$ — $21u^4v + 49u^4 - 14u^3 + 7u^2$
15. $xy^2(3x^2 - xy^2 + 11x - 6y)$ — $3x^3y^2 - x^2y^4 + 11x^2y^2 - 6xy^3$
16. $2d(cd^2 - c^2 + 10cd - 9)$ — $2cd^3 - 2c^2d + 20cd^2 - 18d$
17. A rectangle has a base of length $x^2$ and a height of $3xy^2 - 2x + 1$. Write and simplify an expression for the area of the rectangle.

    $3x^3y^2 - 2x^3 + x^2$




LESSON 14-5

## Practice B

### Multiplying Polynomials by Monomials

Multiply.

1. $(x^2)(-3x^2y^3)$ — $-3x^4y^3$
2. $(-9pr^4)(p^2r^2)$ — $-9p^3r^6$
3. $(2st^9)(-st^2)$ — $-2s^2t^{11}$
4. $(3efg^2)(-3e^2f^2g)$ — $-9e^3f^3g^3$
5. $2q(4q^2 - 2)$ — $8q^3 - 4q$
6. $-x(x^2 + 2)$ — $-x^3 - 2x$
7. $5m(-3m^2 + 2m)$ — $-15m^3 + 10m^2$
8. $6x(-x^5 + 2x^3 + x)$ — $-6x^6 + 12x^4 + 6x^2$
9. $-4st(st - 12t - 2s)$ — $-4s^2t^2 + 48st^2 + 8s^2t$
10. $-9ab(a^2 + 2ab - b^2)$ — $-9a^3b - 18a^2b^2 + 9ab^3$
11. $-7v^2w^2(vw^2 + 2vw + 1)$ — $-7v^3w^4 - 14v^3w^3 - 7v^2w^2$
12. $8p^4(p^2 - 8p + 17)$ — $8p^6 - 64p^5 + 136p^4$
13. $4x(-x^2 - 2xy + 3)$ — $-4x^3 - 8x^2y + 12x$
14. $7x^2(3x^2y + 7x^2 - 2x)$ — $21x^4y + 49x^4 - 14x^3$
15. $-4t^3r^2(3t^2r - t^5r - 6t^2r^2)$ — $-12t^5r^3 + 4t^8r^3 + 24t^5r^4$
16. $h^2k(2hk^2 - hk + 7k)$ — $2h^3k^3 - h^3k^2 + 7h^2k^2$
17. A triangle has a base of $4x^2$ and a height of $6x + 3$. Write and simplify an expression for the area of the triangle.

    $12x^3 + 6x^2$ units²

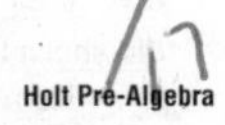




LESSON 14-5

## Practice C

### Multiplying Polynomials by Monomials

Multiply.

1. $(2g^3)(-7g^2)$ — $-14g^5$
2. $(-9q^2r^4)(q^2r^3)$ — $-9q^4r^7$
3. $\left(\frac{1}{3}s^2t^2\right)(3st)$ — $s^3t^3$
4. $\left(\frac{3}{2}g^2\right)\left(-\frac{8}{3}f^2g\right)$ — $-4f^2g^3$
5. $2q(4q^2 - 2q)$ — $8q^3 - 4q^2$
6. $-x^6(x^2 - x + 2)$ — $-x^8 + x^7 - 2x^6$
7. $\frac{1}{5}m^2n\left(5mn^2 - \frac{3}{5}mn\right)$ — $m^3n^3 - \frac{3}{25}m^3n^2$
8. $x^6(-x^5 + 2x^3 + 7)$ — $-x^{11} + 2x^9 + 7x^6$
9. $0.2bc(bc - 0.25c)$ — $0.2b^2c^2 - 0.05bc^2$
10. $-\frac{2}{3}ef\left(e^2 + \frac{1}{2}ef - \frac{3}{5}f^2\right)$ — $-\frac{2}{3}e^3f - \frac{1}{3}e^2f^2 + \frac{2}{5}ef^3$
11. $-7v^2w^2(vw^2 + 2vw - 1)$ — $-7v^3w^4 - 14v^3w^3 + 7v^2w^2$
12. $4x(-x^2 - 2y^2 + 2xy + x + 3)$ — $-4x^3 - 8xy^2 + 8x^2y + 4x^2 + 12x$
13. $2y^2(-y^2z^2 - 2yz^2 + 4y^2 + 2z^2)$ — $-2y^4z^2 - 4y^3z^2 + 8y^4 + 4y^2z^2$
14. $-4h^3k^2(3h^4k - h^2k + 2h^3 + k^6)$ — $-12h^7k^3 + 4h^5k^3 - 8h^6k^2 - 4h^3k^8$
15. A square has a side of length $2y^2$. Write and simplify an expression for the area of the square. Then find the area of the square if $y = 3$.

    $(2y^2)^2 = 4y^4$; 324 square units
16. A rectangle has a base of $3m^2 + m - 2$ and height of $4m^2$. Write and simplify an expression for the area of the rectangle. Then find the area of the rectangle if $m = 2$.

    $12m^4 + 4m^3 - 8m^2$; 192 square units

LESSON **14-5**

## Reteach

### Multiplying Polynomials by Monomials

To multiply a monomial by a monomial, follow the steps used in the example below.

$(7x^2y^3)(3xy^4)$

1. Multiply the coefficients.

   $(7)(3) = 21$

2. Multiply the variables.

   To multiply two powers with the same base, you keep the base and **add** the exponents.

   $(x^2)(x) = (x^2)(x^1) = x^3$ $(y^3)(y^4) = y^7$

3. Write the monomial product.

   $21x^3y^7$

> Remember: If a variable has no exponent, the exponent is 1.
> $x = x^1$

12

**Multiply.**

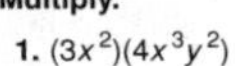

1. $(3x^2)(4x^3y^2)$ — $12x^5y^2$
2. $(6a^3b)(2a^3b^4)$ — $12a^6b^5$
3. $(2m^4n^2)(-5m^2n^2)$ — $-10m^6n^4$

To multiply a polynomial by a monomial, multiply each term of the polynomial by the monomial.

$$\begin{array}{r} 4a^2 + 2ab + 6b^2 \\ \times \quad\quad 3a^3 \\ \hline 12a^5 + 6a^4b + 18a^3b^2 \end{array}$$

**Multiply.**

4. $\begin{array}{r} 3r^2s^3 - 2r^2 + 10 \\ \times \quad 2s \end{array}$ — $6r^2s^4 - 4r^2s + 20s$

5. $\begin{array}{r} 5x^5 + x^2 - 3x \\ \times \quad 4x^3 \end{array}$ — $20x^8 + 4x^5 - 12x^4$

6. $\begin{array}{r} m^2n - 3mn^2 - 8n^3 \\ \times \quad -3mn \end{array}$ — $-3m^3n^2 + 9m^2n^3 + 24mn^4$

16




LESSON **14-5**

## Challenge

### X Marks the Surface

**Write a polynomial for the surface area of each box. Simplify each polynomial. Then use your polynomials to find each box's surface area, if $x = 4$ inches. Remember: For a rectangular prism, $SA = 2lw + 2lh + 2hw$.**

22

1. 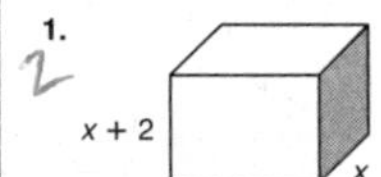


   $10x^2 + 12x$

   SA = $208 \text{ in}^2$

2. 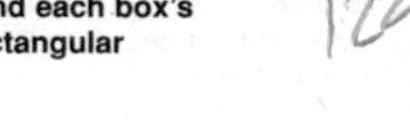


   $14x^2 - 16x$

   SA = $160 \text{ in}^2$

3. 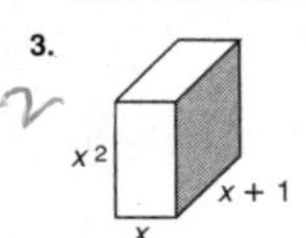


   $4x^3 + 4x^2 + 2x$

   SA = $328 \text{ in}^2$

4. 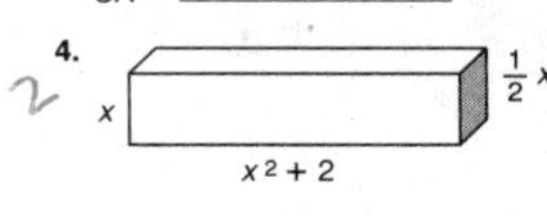


   $3x^3 + x^2 + 6x$

   SA = $232 \text{ in}^2$

5. 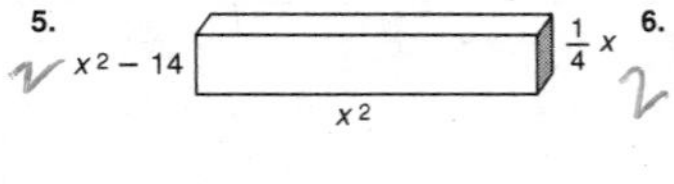


   $2x^4 + x^3 - 28x^2 - 7x$

   SA = $100 \text{ in}^2$

6. 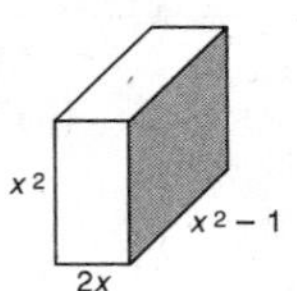


   $2x^4 + 8x^3 - 2x^2 - 4x$

   SA = $976 \text{ in}^2$

LESSON **14-5**

## Problem Solving

### Multiplying Polynomials by Monomials

**Write the correct answer.**

1. A rectangle has a width of $5n^2$ inches and a length of $3n^2 + 2n + 1$ inches. Write and simplify an expression for the area of the rectangle. Then find the area of the rectangle if $n = 2$ inches.

   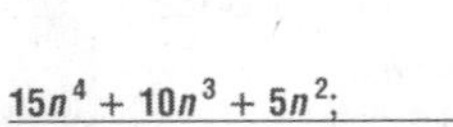

   $15n^4 + 10n^3 + 5n^2$; 340 square inches

2. The area of a parallelogram is found by multiplying the base and the height. Write and simplify an expression for the area of the parallelogram below.

   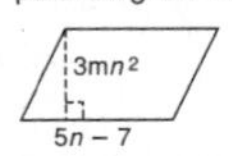


   $15mn^3 - 21mn^2$

3. A parallelogram has a base of $2x^2$ inches and a height of $x^2 + 2x - 1$ inches. Write an expression for the area of the parallelogram. What is the area of the parallelogram if $x = 2$ inches?

   $2x^4 + 4x^3 - 2x^2$; 56 square inches

4. A rectangle has a length of $x^2 + 2x - 1$ meters and a width of $x^2$ meters. Write an expression for the area of the rectangle. What is the area of the rectangle if $x = 3$ meters?

   $x^4 + 2x^3 - x^2$; $126 \text{ m}^2$

**Circle the letter of the correct answer.**

5. A rectangle has a width of $3x$ feet. Its length is $2x + \frac{1}{6}$ feet. Which expression shows the area of the rectangle?

   A $5x + \frac{1}{6}$
   B $6x^2 + \frac{1}{2}x^2$
   C $6x^2 + \frac{1}{2}$
   (D) $6x^2 + \frac{1}{2}x$

6. Which expression shows the area of the shaded region of the drawing?

   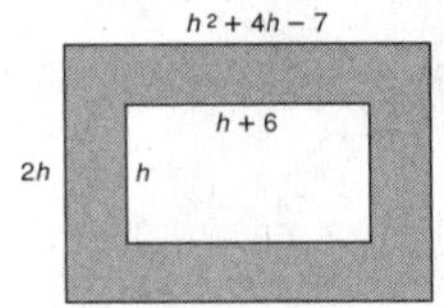


   F $2h^3 + 8h - 14h$
   G $2h^3 + 9h^2 - 8h$
   (H) $2h^3 + 7h^2 - 20h$
   J $2h^3 + 7h^2 - 8h$




LESSON **14-5**

## Reading Strategies

### Draw Conclusions

Follow these rules to multiply polynomials.

| Multiplying Monomials by Monomials | Multiplying Polynomials by Monomials |
|---|---|
| **Rule:** To multiply monomials, multiply the coefficients and add the exponents of the variables that are the same. Remember, if a variable does not have an exponent, the exponent is 1. | **Rule:** To multiply a polynomial by a monomial, use the Distributive Property. Multiply every term of the polynomial by the monomial. |
| **Example:** $(12x^6y^2)(4xy^3)$ <br> $(12x^6y^2)(4xy^3)$ <br> $12 \cdot 4 \cdot x^{6+1}y^{2+3}$ <br> $48x^7y^5$ | **Example:** $-6ab^5(2a^2 + 4ab - 9b)$ <br> $-6ab^5(2a^2 + 4ab - 9b)$ <br> $-12a^3b^5 - 24a^2b^6 + 54ab^6$ |

**Use the information in the chart to answer the following questions.**

1. When multiplying two or more monomials, what do you do to the coefficients of the variables?

   Multiply them.

2. What do you do to the exponents of the variables that are the same?

   Add them.

3. If a variable does not have an exponent, what is the exponent?

   1

4. Write $y^2$ times $y^3$ without using exponents to show why you add the exponents when multiplying.

   $(yy)(yyy) = y^5$

5. What property do you use when multiplying a polynomial by a monomial?

   the Distributive Property

6. How many terms will you get if you multiply $2x^2y$ by $(4xy^3 + 2x - 5)$?

   You will get 3 terms.

16

LESSON 14-5
## Puzzles, Twisters & Teasers
### Tic–Tac–Nomials!

**Simplify the expressions in each square to determine whether they are equal. If they are equal, place an X in the square. If they are not equal, place an O in the square. Be sure to mark the Tic–Tac–Toe!**

| | | |
|---|---|---|
| $(12x^2y)(2x^3y^4)$ ? = $(6x)(4x^4y)$ **O** | $2a^2b^2(20a + 8ab - 6b^2)$ ? = $4b^2(10a^3 + 4a^3b - 3a^2b^2)$ **X** | $-6m(4 + 8m^2)$ ? = $2m(-12 + 24m^2)$ **O** |
| $-2c(4c^2d) + 10c^3d$ ? = $2c(c^2d)$ **X** | $(3rs^3)(-8r^4s)$ ? = $(-6r^5s^2)(4s^2)$ **X** | $-5x(4x^2y - 6xy + 8y^2)$ ? = $10y(3x^2 - 2x^3 - 4xy)$ **X** |
| $-4gh^2(9g^2 + 6h^2)$ ? = $12g(-2h^4 - 3g^2h^2)$ **X** | $6a(a^2 - 10a) + 50a^2$ ? = $2a^2(3a + 5)$ **O** | $(12xy)(-6x^3y)$ ? = $(-8x^2y^2)(-9x^2)$ **O** |




LESSON 14-6
## Practice A
### Multiplying Binomials

**Multiply.**

1. $(z + 3)(z + 2)$ $z^2 + 5z + 6$
2. $(a + 1)(a + 2)$ $a^2 + 3a + 2$
3. $(b + 1)(3b + 1)$ $3b^2 + 4b + 1$
4. $(c + 1)(c - 3)$ $c^2 - 2c - 3$
5. $(3v - 1)(3v - 3)$ $9v^2 - 12v + 3$
6. $(2r + 1)(2r - 1)$ $4r^2 - 1$
7. $(u - 5v)(3u - v)$ $3u^2 - 16uv + 5v^2$
8. $(-3g + 4)(g - 1)$ $-3g^2 + 7g - 4$
9. $(-2r + s)(r + 6s)$ $-2r^2 - 11rs + 6s^2$
10. Write an expression for the area of a cement walkway of width $k$ ft around an 8 ft by 10 ft jacuzzi.
    $4k^2 + 36k$ ft$^2$
11. Write and simplify an expression for the area of a brick walkway of width $x$ m around a 30 m by 50 m yard.
    $4x^2 + 160x$ m$^2$
12. A rug is placed in a 15 ft × 25 ft room so that there is an uncovered strip of width $h$ feet all the way around the rug. Write and simplify an expression for the area of the rug.
    $4h^2 - 80h + 375$ ft$^2$

**Multiply.**

13. $(e + 1)^2$ $e^2 + 2e + 1$
14. $(2a + 1)^2$ $4a^2 + 4a + 1$
15. $(e - 3)^2$ $e^2 - 6e + 9$
16. $(m - 1)(m + 1)$ $m^2 - 1$
17. $(2 + e)^2$ $4 + 4e + e^2$
18. $(k - n)^2$ $k^2 - 2kn + n^2$
19. $(2w - z)(2w + z)$ $4w^2 - z^2$
20. $(2a + b)(2a - b)$ $4a^2 - b^2$




LESSON 14-6
## Practice B
### Multiplying Binomials

**Multiply.**

1. $(z + 1)(z + 2)$ $z^2 + 3z + 2$
2. $(1 - y)(2 - y)$ $2 - 3y + y^2$
3. $(2x + 1)(2x + 4)$ $4x^2 + 10x + 4$
4. $(w + 1)(w - 3)$ $w^2 - 2w - 3$
5. $(3v + 1)(v - 1)$ $3v^2 - 2v - 1$
6. $(t + 2)(2t - 2)$ $2t^2 + 2t - 4$
7. $(-3g + 4)(2g - 1)$ $-6g^2 + 11g - 4$
8. $(3c + d)(c - 2d)$ $3c^2 - 5cd - 2d^2$
9. $(2a + b)(a + 2b)$ $2a^2 + 5ab + 2b^2$
10. A box is formed from a 1 in. by 18 in. piece of cardboard by cutting a square with side length $m$ inches out of each corner and folding up the sides. Write and simplify an expression for the area of the base of the box.
    $18 - 38m + 4m^2$ in.$^2$
11. A table is placed in a 14 ft × 18 ft room so that there is an equal amount of space of width $s$ feet all the way around the table. Write and simplify an expression for the area of the table.
    $252 - 64s + 4s^2$ ft$^2$
12. A circular swimming pool with a radius of 14 ft is surrounded by a deck with width $y$ feet. Write and simplify an expression for the total area of the pool and the deck. Use $\frac{22}{7}$ for pi.
    $616 + 88y + \frac{22}{7}y^2$ ft$^2$

**Multiply.**

13. $(r - 2)^2$ $r^2 - 4r + 4$
14. $(2 + q)^2$ $4 + 4q + q^2$
15. $(p + 4)(p - 4)$ $p^2 - 16$
16. $(3n - 3)(3n + 3)$ $9n^2 - 9$
17. $(a + b)(a - b)$ $a^2 - b^2$
18. $(4e - f)^2$ $16e^2 - 8ef + f^2$
19. $(2y + z)^2$ $4y^2 + 4yz + z^2$
20. $(9p - 2)(-2 + 9p)$ $81p^2 - 36p + 4$
21. $(m - 1)^2$ $m^2 - 2m + 1$




LESSON 14-6
## Practice C
### Multiplying Binomials

**Multiply.**

1. $(a + 1)(a + 2)$ $a^2 + 3a + 2$
2. $(1 - x)(2 - x)$ $2 - 3x + x^2$
3. $(2c + 1)(2c + 4)$ $4c^2 + 10c + 4$
4. $(2w + 1)(-2w - 3)$ $-4w^2 - 8w - 3$
5. $\left(2v + \frac{1}{4}\right)\left(v - \frac{1}{4}\right)$ $2v^2 - \frac{1}{4}v - \frac{1}{16}$
6. $(3t + 2)(2t - 3)$ $6t^2 - 5t - 6$
7. A circular swimming pool is surrounded by a deck that is $y$ meters wide. The radius of the swimming pool is 21 meters. Write and simplify an expression for the total area of the swimming pool and deck. Use $\frac{22}{7}$ for pi.
   $\frac{22}{7}y^2 + 132y + 1{,}386$ m$^2$
8. A rug is placed in a 6 ft × 15 ft hallway so that there is an equal amount of space of width $y$ feet all the way around the rug. Write and simplify an expression for the area of the rug.
   $90 - 42y + 4y^2$ ft$^2$
9. A brick walkway around a circular jacuzzi is 7 ft wide. The radius of the jacuzzi is $14d$ ft. Write and simplify an expression for the total area of the jacuzzi and the brick walkway around the jacuzzi. Use $\frac{22}{7}$ for pi.
   $616d^2 + 616d + 154$ ft$^2$

**Multiply.**

10. $(p + 2)^2$ $p^2 + 4p + 4$
11. $(2 - k)^2$ $4 - 4k + k^2$
12. $(d + 14)(d - 14)$ $d^2 - 196$
13. $(3c - 3d)(3c + 3d)$ $9c^2 - 9d^2$
14. $(2a + b)(2a - b)$ $4a^2 - b^2$
15. $\left(\frac{1}{2}c + d\right)\left(\frac{1}{2}c - d\right)$ $\frac{1}{4}c^2 - d^2$
16. $(9p - 2q)(-2q + 9p)$ $81p^2 - 36pq + 4q^2$
17. $(2m - 3b)^2$ $4m^2 - 12bm + 9b^2$
18. $(9a + 3b)(9a - 3b)$ $81a^2 - 9b^2$

## LESSON 14-6 Reteach
### *Multiplying Binomials*

To multiply a binomial by a binomial, multiply each term of the first binomial by each term of the second binomial.

$$(a + b)(c + d) = ac + ad + bc + bd$$

You can remember the product as FOIL: First terms, Outer terms, Inner terms, and Last terms.

$(5x + 3)(3x - 2)$

| | |
|---|---|
| Multiply the **F**irst terms. | $(5x)(3x) = 15x^2$ |
| Multiply the **O**utside terms. | $(5x)(-2) = -10x$ |
| Multiply the **I**nside terms. | $(3)(3x) = 9x$ |
| Multiply the **L**ast terms. | $(3)(-2) = -6$ |
| Add the products. | $15x^2 - 10x + 9x - 6$ |
| Combine like terms. | $15x^2 - x - 6$ |

**You can also multiply binomials vertically.**

| | |
|---|---|
| Align the binomials. | $5x + 3$ |
| Multiply each term of one | $\times\ 3x - 2$ |
| binomial by each term of the | $-10x - 6$ |
| other binomial. | $15x^2 + 9x$ |
| Combine like terms. | $15x^2 - x - 6$ |

**Multiply.**

1. $(4x + 3)(2x + 5)$ $\underline{8x^2 + 26x + 15}$
2. $(7t - 4)(2t + 3)$ $\underline{14t^2 + 13t - 12}$
3. $(3 + 5b)(2b - 3b^2)$ $\underline{6b + b^2 - 15b^3}$
4. $(x - 1)(x + 5)$ $\underline{x^2 + 4x - 5}$
5. $(6m - 3n)(2m + 3n)$ $\underline{12m^2 + 12mn - 9n^2}$
6. $(c + 7)(c + 1)$ $\underline{c^2 + 8c + 7}$
7. $6n - 3$ $\times\ 3n + 3$ $\underline{18n^2 + 9n - 9}$
8. $2y + 4$ $\times\ y + 6$ $\underline{2y^2 + 16y + 24}$




## LESSON 14-6 Challenge
### *Multiplication Tables*

You can use a table to multiply binomials. Label each column with a term from one of the binomials, and label each row with a term from the other binomial of the coefficients. Then multiply the same as in a multiplication table (column × row). Finish by combining like terms in the product.

**Example: $(x + 3)(x - 2)$**

| | $x$ | $3$ |
|---|---|---|
| $x$ | $x^2$ | $3x$ |
| $-2$ | $-2x$ | $-6$ |

$x^2 + 3x - 2x - 6 = \mathbf{x^2 + x - 6}$

**Use the given tables to find each product.**

1. $(x - 4)(x - 5)$

| | $x$ | $-4$ |
|---|---|---|
| $x$ | $x^2$ | $-4x$ |
| $-5$ | $-5x$ | $20$ |

$\underline{x^2 - 9x + 20}$

2. $(3m + 5)^2$

| | $3m$ | $5$ |
|---|---|---|
| $3m$ | $9m^2$ | $15m$ |
| $5$ | $15m$ | $25$ |

$\underline{9m^2 + 30m + 25}$

3. $(2h + 1)(h - 4)$

| | $2h$ | $1$ |
|---|---|---|
| $h$ | $2h^2$ | $h$ |
| $-4$ | $-8h$ | $-4$ |

$\underline{2h^2 - 7h - 4}$

4. $(7p - q)(p + 6q)$

| | $7p$ | $-q$ |
|---|---|---|
| $p$ | $7p^2$ | $-pq$ |
| $6q$ | $42pq$ | $-6q^2$ |

$\underline{7p^2 + 41pq - 6q^2}$

5. $(3a + 4b)^2$

| | $3a$ | $4b$ |
|---|---|---|
| $3a$ | $9a^2$ | $12ab$ |
| $4b$ | $12ab$ | $16b^2$ |

$\underline{9a^2 + 24ab + 16b^2}$

6. $(5w - 10)(2w - v)$

| | $5w$ | $-10$ |
|---|---|---|
| $2w$ | $10w^2$ | $-20w$ |
| $-v$ | $-5wv$ | $10v$ |

$\underline{10w^2 - 5wv - 20w + 10v}$




## LESSON 14-6 Problem Solving
### *Multiplying Binomials*

**Write and simplify an expression for the area of each polygon.**

| | Polygon | Dimensions | Area |
|---|---|---|---|
| 1. | rectangle | length: $(n + 5)$; width: $(n - 4)$ | $n^2 + n - 20$ |
| 2. | rectangle | length: $(3y + 3)$; width: $(2y - 1)$ | $6y^2 + 3y - 3$ |
| 3. | triangle | base: $(2b - 5)$; height: $(b^2 + 2)$ | $b^3 - \frac{5}{2}b^2 + 2b - 5$ |
| 4. | square | side length: $(m + 13)$ | $m^2 + 26m + 169$ |
| 5. | square | side length: $(2g - 4)$ | $4g^2 - 16g + 16$ |
| 6. | circle | radius: $(3c + 2)$ | $(9c^2 + 12c + 4)\pi$ |

**Choose the letter of the correct answer.**

7. A photo is 8 inches by 11 inches. A frame of width $x$ inches is placed around the photo. Which expression shows the total area of the frame and photo?
   - A $x^2 + 19x + 88$
   - Ⓑ $4x^2 + 38x + 88$
   - C $8x + 38$
   - D $4x + 19$

8. Three consecutive odd integers are represented by the expressions, $x$, $(x + 2)$ and $(x + 4)$. Which expression gives the product of the three odd integers?
   - F $x^3 + 8$
   - Ⓖ $x^3 + 6x^2 + 8x$
   - H $x^3 + 6x^2 + 8$
   - J $x^3 + 2x^2 + 8x$

9. A square garden has a side length of $(b - 4)$ yards. Which expression shows the area of the garden?
   - A $2b - 8$
   - B $b^2 + 16$
   - C $b^2 - 8b - 16$
   - Ⓓ $b^2 - 8b + 16$

10. Which expression gives the product of $(3m + 4)$ and $(9m - 2)$?
   - Ⓕ $27m^2 + 30m - 8$
   - G $27m^2 + 42m - 8$
   - H $27m^2 + 42m + 8$
   - J $27m^2 + 30m + 8$




## LESSON 14-6 Reading Strategies
### *Use Graphic Aids*

This chart can help you remember how to multiply binomials.

| How to Multiply Binomials | Special Products of Binomials |
|---|---|
| FIRST, OUTER, INNER, LAST<br>First Last<br>$(a + b)(c + d)$<br>Inner<br>Outer<br>$ac + ad + bc + bd$<br>The Commutative Property states that it does not matter in which order you write the addends. | $\mathbf{(a + b)^2} = (a + b)(a + b) = a^2 + 2ab + b^2$<br>**Example:** $(x + 4)^2 = x^2 + 2(4)(x) + 4^2 = x^2 + 8x + 16$<br>$\mathbf{(a - b)^2} = (a - b)(a - b) = a^2 - 2ab + b^2$<br>**Example:** $(x - 6)^2 = x^2 - 2(6)(x) + 6^2 = x^2 - 12x + 36$<br>$\mathbf{(a + b)(a - b)} = a^2 - b^2$<br>**Example:** $(x + 7)(x - 7) = x^2 - 7^2 = x^2 - 49$ |

**Use the information in the chart to answer the following questions.**

1. What do the letters FOIL represent?

   first, outer, inner, last

2. Does it matter in which order you write the addends? Why or why not??

   No; the Commutative Property guarantees that $ac + ad = ad + ac$, etc.

3. When finding the special product of $(a + b)^2$, how many terms will you always have?

   three

4. When finding the special product of $(a - b)^2$, why is the last term positive?

   because negative $b$ times negative $b$ is positive $b^2$

5. Why do you get only two terms when you multiply binomials in the special form $(a + b)(a - b)$?

   The middle two terms are $ab$ and $-ab$, which cancel each other out.

LESSON **14-6**

## Puzzles, Twisters & Teasers

### *Back It Up!*

**What do you have when a row of rabbits steps backwards?**

**To find the answer, find each product in Column 1 and match it to the correct expression in Column 2. Then, write the letter above the corresponding exercise number.**

| Column 1 | | Column 2 |
|---|---|---|
| **1.** $(x + 4)^2$ | L | **A** $x^2 + 4x - 32$ |
| **2.** $(x + 3)(x + 6)$ | C | **C** $x^2 + 9x + 18$ |
| **3.** $(x + 8)(x - 4)$ | A | **D** $x^2 - 16$ |
| **4.** $(x + 4)(x - 4)$ | D | **E** $6x^2 - 2x - 20$ |
| **5.** $(x - 9)(x - 2)$ | N | **G** $9x^2 - 30x + 25$ |
| **6.** $(x - 16)(x + 2)$ | H | **H** $x^2 - 14x - 32$ |
| **7.** $(2x - 5)(3x + 4)$ | I | **I** $6x^2 - 7x - 20$ |
| **8.** $(6x + 10)(x - 2)$ | E | **L** $x^2 + 8x + 16$ |
| **9.** $(3x - 5)^2$ | G | **N** $x^2 - 11x + 18$ |
| **10.** $(3x - 5)(3x + 5)$ | R | **R** $9x^2 - 25$ |

| A | R | E | C | E | D | I | N | G |
|---|---|---|---|---|---|---|---|---|
| 3 | 10 | 8 | 2 | 8 | 4 | 7 | 5 | 9 |

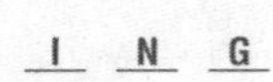

| H | A | R | E | L | I | N | E |
|---|---|---|---|---|---|---|---|
| 6 | 3 | 10 | 8 | 1 | 7 | 5 | 8 |